Multilevel Statistical Models

Second Edition

KENDALL'S ADVANCED THEORY OF STATISTICS
and
KENDALL'S LIBRARY OF STATISTICS

Advisory Editorial Board: PJ Green, University of Bristol; RJ Little, University of Michigan; JK Ord, Pennsylvania State University; S Weisberg, University of Minnesota

The development of statistical theory in the past fifty years is faithfully reflected in the history of the late Sir Maurice Kendall's volumes THE ADVANCED THEORY OF STATISTICS. The ADVANCED THEORY began life as a two volume work (Volume 1, 1943; Volume 2, 1946) and grew steadily, as a single authored work, until the late fifties. At that point, Alan Stuart became co-author and the ADVANCED THEORY was rewritten in three volumes. When Keith Ord joined in the early eighties, Volume 3 became the largest and plans were developed to expand, yet again, to a four-volume work. Even so, it became evident that there were gaps in the coverage and that it was becoming increasingly difficult to provide timely updates to all volumes, so a new strategy was devised.

In future, the ADVANCED THEORY will be in the form of three core volumes together with a series of related monographs called KENDALL'S LIBRARY OF STATISTICS. The three volumes of ADVANCED THEORY will be:

1   Distribution Theory
2A   Classical Inference and Relationships
2B   Bayesian Inference (a new companion volume by Anthony O'Hagan)

KENDALL'S LIBRARY OF STATISTICS will encompass the areas previously appearing in the old Volume 3, such as sample surveys, design of experiments, multi-variate analysis and time series as well as non-parametrics and log-linear models, previously covered to some extent in Volume 2. In the preface to the first edition of THE ADVANCED THEORY Kendall declared that his aim was 'to develop a systematic treatment of [statistical] theory as it exists at the present time' while ensuring that the work remained 'a book on statistics, not on statistical mathematics'. These aims continue to hold true for KENDALL'S LIBRARY OF STATISTICS and the flexibility of the monograph format will enable the series to maintain comprehensive coverage over the whole of modern statistics.

Published volumes: 1. MULTIVARIATE ANALYSIS Part 1 Distibutions, Ordination and
Inference, WJ Krzanowski (University of Exeter) and
FHC Marriott (University of Oxford)   1994   0 340 59326 1

2. MULTIVARIATE ANALYSIS Part 2 Classification, Covariance Structures
and Repeated Measurements, WJ Krzanowski (University of Exeter) and
FHC Marriott (University of Oxford)   1995   0 340 59325 3

3. MULTILEVEL STATISTICAL MODELS Second Edition,
Harvey Goldstein (University of London)   1995   0 340 59529 9

MULTILEVEL STATISTICAL MODELS

Second Edition

Harvey Goldstein

Institute of Education, University of London

A member of the Hodder Headline Group
LONDON • SYDNEY • AUCKLAND

Copublished in the Americas by Halsted Press
an imprint of John Wiley & Sons Inc.
New York – Toronto

First published in Great Britain 1987 as Multilevel Models in Educational
and Social Research
Second edition published 1995 as Multilevel Statistical Models by Edward Arnold
Reprinted 1996 by Arnold, a member of the Hodder Headline Group
338 Euston Road, London NW1 3BH

Copublished in the Americas by Halsted Press, an imprint of
John Wiley & Sons Inc., 605 Third Avenue, New York, NY 10158

British Library Cataloguing in Publication Data
A catalogue record for this book is available from the British Library

ISBN 0 340 59529 9

Library of Congress Cataloging-in-Publication Data
A catalog record for this book is available from the Library of Congress

ISBN 0 470 24989 7 (In the Americas only)

Typeset in 10/11pt Times by Wearset, Boldon, Tyne and Wear
Printed and bound in Great Britain by St Edmundsbury Press,
Bury St Edmunds, Suffolk and Bookcraft, Bath, Avon.

Contents

Preface

In the mid 1980s a number of researchers began to see how to introduce systematic approaches to the statistical modelling and analysis of hierarchically structured data. The early work of Aitkin *et al* (1981) on the teaching styles' data and Aitkin's subsequent classic work with Longford (1986) initiated a series of developments that, by the early 1990s, had resulted in a core set of established techniques, experience and software packages that could be applied routinely. These methods and further extensions of them are described in this book and are coming to be applied widely in areas such as education, epidemiology, geography, child growth, household surveys and many others.

In addition to the first edition of the present text (Goldstein, 1987b), two expository volumes appeared in the early 1990s. That by Bryk and Raudenbush (1992) discussed 2- and 3-level linear multilevel models with applications to educational data in particular and also to repeated measures designs. Longford (1993) gave a more theoretically oriented account and additionally included discussion of a multilevel factor analysis model, models with categorical responses and multivariate models. The present volume aims to integrate existing methodological developments within a consistent terminology and notation, provide examples and explain a number of new developments, especially in the areas of discrete response data, time series models, random cross classifications, errors of measurement, missing data and nonlinear models. In many cases these developments are the subject of continuing research, so that we can expect further elaborations of the procedures described.

The main text seeks to avoid undue statistical complexity, with methodological derivations occurring in appendices. Examples and diagrams are used to illustrate the application of the techniques and references are given to other work. The book is intended to be suitable for graduate level courses and as a general reference.

Harvey Goldstein
London

Acknowledgements

This book would not have been possible without the support and dedication of all those who have worked on or been closely associated with the Multilevel Models Project at the Institute of Education: Jon Rasbash, Min Yang, Bob Prosser, Geoff Woodhouse, Pan Huiqui, Michael Healy and Ian Plewis. I am also most grateful to the Economic and Social Research Council for their continuing funding support since 1986, most recently under the auspices of the programme for the Analysis of Large and Complex Datasets.

In addition, the following, most generously, have allowed me access to datasets or commented critically on chapter drafts: Bob Carpenter, Arto Demirjian, David Draper, Peter Egger, Anthony Heath, Kelvyn Jones, Ita Kreft, Toby Lewis, Mac MacDonald, Rod McDonald, Colm O'Muircheartaigh, Lindsay Paterson, Steve Raudenbush, German Rodriguez, Pamela Sammons, Richard Tanguay and Sally Thomas.

Notation

The following definitions refer to a 2-level model. The extension to three and higher level models is usually straightforward. Where this is not clear, a three-level definition is included.

Definition	Symbol
Response variable vector	Y
Explanatory variable design matrix	X
Fixed part explanatory variable design vector for a single unit	X_{ij} for a level 1 unit X_j for a level 2 unit
Total residuals at each level for a 3-level model	$v_k = \sum_{h=0}^{q_3} v_{hk} z_{hk}^{(3)}$ $u_{jk} = \sum_{h=0}^{q_2} u_{hjk} z_{hjk}^{(2)}$ $e_{ijk} = \sum_{h=0}^{q_1} e_{hijk} z_{hijk}^{(1)}$
Explanatory variable design matrix for level 2 and level 1 random coefficients	$Z^{(2)}, Z^{(1)}$
Predicted value from fixed part of model	$\hat{y}_{ij} = X_{ij}\beta = (X\beta)_{ij}$
Raw or total residual for level 1 unit	$\tilde{y}_{ij} = y_{ij} - \hat{y}_{ij}$
Mean raw residual for level 2 unit	$\tilde{y}_j = \dfrac{1}{n_j}\sum_{i=1}^{n_j} \tilde{y}_{ij}$

Estimated residual or posterior residual estimate	$\hat{u}_j$, $\hat{e}_{ij}$
Covariance matrix of random coefficients at level i	Ω_i
Parentheses denoting vector or matrix of elements	{}
Covariance matrix of response vector for k-level model	V_k or just V
Contribution to covariance matrix of response vector from level i for k-level model	$V_{k(i)}$, or just $V_{(i)}$
Direct sum of matrices A_1, ..., A_k	$\bigoplus_{i=1}^{k} A_i$
Kronecker product of conformable matrices A_1, A_2	$A_1 \otimes A_2$
vec operator on matrix A	$\text{vec}(A)$

Glossary

Cluster	A grouping containing 'lower level' elements. For example, in a sample survey the set of households in a neighbourhood.
Design matrix	In the fixed part of the model, the matrix of values of the explanatory variables X. In the random part the matrix of explanatory variables Z.
Explanatory variable	Also known as an 'independent' variable. In the fixed part of the model usually denoted by x and in the random part by z.
Fixed part	That part of a model represented by $X\beta$, that is the average relationship.
Level	A component of a data hierarchy. Level 1 is the lowest level, for example students within schools or repeated measurement occasions within individual subjects.
Level n variation	The variation of level n unit measurements about the fixed part of a model.
Nesting	The clustering of units into a hierarchy
Random part	That part of a model represented by Zu; that is, the contribution of the random variables at each level.
Response variable	Also known as a 'dependent' variable. Denoted by y.

Unit

An entity defined at a level of a data hierarchy. For example, an individual student will be a level 1 unit within a level 2 unit which is a school.

1

Introduction

Introduction

1.1 Many kinds of data, including observational data collected in the human and biological sciences, have a *hierarchical* or *clustered* structure. For example, animal and human studies of inheritance deal with a natural hierarchy where offspring are grouped within families. Offspring from the same parents tend to be more alike in their physical and mental characteristics than individuals chosen at random from the population at large. For instance, children from the same family may all tend to be small, perhaps because their parents are small or because of a common impoverished environment.

Many designed experiments also create data hierarchies; for example, clinical trials carried out in several randomly chosen centres or groups of individuals. For now, we are concerned only with the *fact* of such hierarchies, not their provenance. The principal applications we shall deal with are those from the social sciences, but the techniques are, of course, applicable more generally. In subsequent chapters, as we develop the theory and techniques with examples, we shall see how a proper recognition of these natural hierarchies allows us to seek more satisfactory answers to important questions.

We refer to a hierarchy as consisting of *units* grouped at different *levels*. Thus, offspring may be the level 1 units in a 2-level structure where the level 2 units are the families: students may be the level 1 units clustered within schools that are the level 2 units.

The existence of such data hierarchies is neither accidental nor ignorable. Individual people differ as do individual animals and this necessary differentiation is mirrored in all kinds of social activity where the latter is often a direct result of the former, for example when students with similar motivations or aptitudes are grouped in highly selective schools or colleges. In other cases, the groupings may arise for reasons less strongly associated with the characteristics of individuals, such as the allocation of young children to elementary schools, or the allocation of patients to different clinics. Once groupings are established, even if their establishment is effectively random, they will tend to become differentiated, and this differentiation implies that the group and its members both influence and are influenced by the group membership. To ignore this relationship risks overlooking the importance of group effects, and may also render invalid many of the traditional statistical analysis techniques used for studying data relationships.

We shall be looking at this issue of statistical validity in the next chapter, but one simple example will show its importance. A well-known and influential study of primary (elementary) school children carried out in the 1970s (Bennett, 1976) claimed that children exposed to so-called 'formal' styles of teaching reading exhibited more progress than those who were not. The data were analysed using traditional multiple regression techniques which recognized only the individual children as the units of analysis and ignored their groupings within teachers and into classes. The results were statistically significant. Subsequently, Aitkin *et al* (1981) demonstrated that when the analysis accounted properly for the grouping of children into classes, the significant differences disappeared and the 'formally' taught children could not be shown to differ from the others.

This re-analysis was the first important example of a *multilevel* analysis of social science data. In essence, what was occurring here was that the children within any one classroom, because they were taught together, tended to be similar in their performance. As a result, they provided rather less information than would have been the case if the same number of students had been taught separately by different teachers. In other words, the basic unit for comparative purposes should have been the teacher, not the student. The function of the students can be seen as providing, for each teacher, an estimate of that teacher's effectiveness. Increasing the number of students per teacher would increase the precision of those estimates but not change the number of teachers being compared. Beyond a certain point, simply increasing the numbers of students in this way hardly improves things at all. On the other hand, increasing the number of teachers to be compared, with the same or a somewhat smaller number of students per teacher, considerably improves the precision of the comparisons.

Researchers have long recognized this issue. In education, for example, there has been much debate (see Burstein *et al*, 1980) about the so-called 'unit of analysis' problem, which is the one just outlined. Before multilevel modelling became well developed as a research tool, the problems of ignoring hierarchical structures were reasonably well understood, but they were difficult to solve because powerful general purpose tools were unavailable. Special purpose software—for example for the analysis of genetic data—has been available longer but this was restricted to 'variance components' models (see Chapter 2) and was not suitable for handling general linear models. Sample survey workers have recognized this issue in another form. When population surveys are carried out, the sample design typically mirrors the hierarchical population structure, in terms of geography and household membership. Elaborate procedures have been developed to take such structures into account when carrying out statistical analyses. We return to this in a little more detail in a later section.

In the remainder of this chapter we shall look at the major areas explored in this book.

School effectiveness

1.2   Schooling systems present an obvious example of a hierarchical structure, with pupils grouped or nested or clustered within schools, which themselves may be clustered within education authorities or boards. Educational researchers have been interested in comparing schools and other educational institutions, most often in terms of the achievements of their pupils. Such comparisons have several aims, including the aim of public accountability (Goldstein, 1992) but, in research terms, interest is usually focused upon studying the factors that explain school differences.

Consider the common example where test or examination results at the end of a period of schooling are collected for each school in a randomly chosen sample of schools. The researcher wants to know whether a particular kind of subject streaming practice in some schools is associated with improved examination performance. The researcher also has good measures of the pupils' achievements when they started the period of schooling so that he or she can control for this in the analysis. The traditional approach to the analysis of these data would be to carry out a regression analysis, using performance score as response, to study the relationship with streaming practice, adjusting for the initial achievements. This is very similar to the initial teaching styles analysis described in the previous section, and suffers from the same lack of validity through failing to take account of the school level clustering of students.

An analysis that explicitly models the manner in which students are grouped within schools has several advantages. First, it enables data analysts to obtain statistically efficient estimates of regression coefficients. Secondly, by using the clustering information it provides correct standard errors, confidence intervals and significance tests, and these generally will be more 'conservative' than the traditional ones, which are obtained simply by ignoring the presence of clustering—just as Bennett's previously statistically significant results became non-significant on re-analysis. Thirdly, by allowing the use of covariates measured at any of the levels of a hierarchy, it enables the researcher to explore the extent to which differences in average examination results between schools are accountable for by factors such as organizational practice or possibly in terms of other characteristics of the students. It also makes it possible to study the extent to which schools differ for different kinds of students, for example to see whether the variation between schools is greater for initially high scoring students than for initially low scoring students (Goldstein *et al*, 1993) and whether some factors are better at accounting for or 'explaining' the variation for the former students than for the latter. Finally, there is often considerable interest in the relative ranking of individual schools, using the performances of their students after adjusting for intake achievements. This can be done straightforwardly using a multilevel modelling approach.

To fix the basic notion of a level and a unit, consider Figs 1.1 and 1.2 based on hypothetical relationships.

Figure 1.1 shows the exam score and intake achievement scores for five students in a school, together with a simple regression line fitted to the data points. The residual variation in the exam scores about this line is the *level 1 residual variation*, since it relates to level 1 units (students) within a sample level 2 unit (school). In Fig. 1.2 the three lines are the simple regression lines for three schools, with the individual student data points removed. These vary in both their slopes and their in-

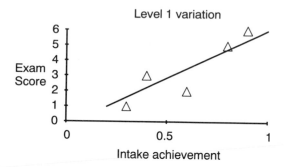

Figure 1.1   Examination score by intake achievement for five students in a school.

tercepts (where they would cross the exam axis), and this variation is *level 2 varia-tion*. It is an example of multiple or complex level 2 variation since both the inter-cept and slope parameters vary.

The other extreme to an analysis which ignores the hierarchical structure, is one which treats each school completely separately by fitting a different regression model within each one. In some circumstances, for example where we have very few schools and moderately large numbers of students in each, this may be efficient. It may also be appropriate if we are interested in making inferences about just those schools. If, however, we regard these schools as a (random) sample from a population of schools and we wish to make inferences about the variation be-tween schools in general, then a full multilevel approach is called for. Likewise, if some of our schools have very few students, fitting a separate model for each of these will not yield reliable estimates: we can obtain more precision by regarding the schools as a sample from a population and using the information available from the whole sample data when making estimates for any one school. This approach is especially important in the case of repeated-measures data, where we typically have very few level 1 units per level 2 unit.

We introduce the basic procedures for fitting multilevel models to hierarchically structured data in Chapter 2 and discuss the design problem of choosing the num-bers of units at each level in Chapter 11.

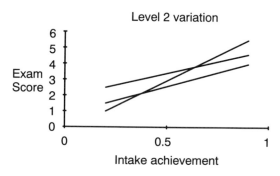

Figure 1.2   Average examination score by intake achievement for three schools.

Sample survey methods

1.3   We have already mentioned sample survey data, which will be discussed in many of the examples of this book. The standard literature on surveys, reflected in survey practice, recognizes the importance of taking account of the clustering in complex sample designs. Thus, in a household survey, the first stage sampling unit will often be a well-defined geographical unit. From those which are randomly chosen, further stages of random selection are carried out until the final households are selected. Because of the geographical clustering exhibited by measures such as political attitudes, special procedures have been developed to produce valid statistical inferences, for example when comparing mean values or fitting regression models (Skinner *et al*, 1989).

While such procedures have usually been regarded as necessary they have not generally merited serious substantive interest. In other words, the population structure, insofar as it is mirrored in the sampling design, is seen as a 'nuisance factor'. By contrast, the multilevel modelling approach views the population structure as of potential interest in itself, so that a sample designed to reflect that structure is not merely a matter of saving costs as in traditional survey design, but can be used to collect and analyse data about the higher level units in the population. The subsequent modelling can then incorporate this information and obviate the need to carry out special adjustment procedures, which are built into the analysis model directly.

Although the direct modelling of clustered data is statistically efficient, it will generally be important to incorporate weightings in the analysis that reflect the sample design or, for example, patterns of non-response, so that robust population estimates can be obtained and so that there will be some protection against serious model misspecification. One procedure for introducing external unit weights into a multilevel analysis is discussed in Chapter 3.

Repeated measures data

1.4   A different example of hierarchically structured data occurs when the same individuals or units are measured on more than one occasion. A common example occurs in studies of animal and human growth. Here, the occasions are clustered within individuals that represent the level 2 units with measurement occasions being the level 1 units. Such structures are typically strong hierarchies because there is much more variation between individuals in general than between occasions within individuals. In the case of child height growth, for example, once we have adjusted for the overall trend with age, the variance between successive measurements on the same individual is generally no more than 5% of the variation in height between children.

There is a considerable past literature on procedures for the analysis of such repeated measurement data (see for example Goldstein, 1979), which has more or less successfully confronted the statistical problems. It has done so at the cost, however, of requiring that the data conform to a particular, balanced, structure. Broadly speaking, these procedures require that the measurement occasions are the same for each individual. This may be possible to arrange, but often, in practice, individuals will be measured irregularly, some of them a great number of times and some perhaps only once. By considering such data as a general 2-level structure we can apply the standard set of multilevel modelling techniques that allow any pattern of measurements while providing statistically efficient parameter estimation. At the same time, modelling a 2-level structure presents a simpler conceptual understanding of such data and leads to a number of interesting extensions that will be explored in Chapter 6.

One particularly important extension occurs in the study of growth where the aim is to fit growth curves to measurements over time. In a multilevel framework this involves, in the simplest case, each individual having their own straight line growth trajectory with the intercept and slope coefficients varying between individuals (level 2). When the level 1 measurements, considered as deviations from each individual's fitted growth curve, are not independent but have an autocorrelated or time series structure, neither the traditional procedures nor the basic multilevel ones are adequate. This situation may occur, for example, when measurements are made very close together in time so that a 'positive' deviation from the curve at one time implies a positive deviation after the short interval before the next measurement.

Event history models

1.5   Modelling time spent in various states or situations is important in a number of areas. In industry, the 'time to failure' of components is a key factor in quality control. In medicine, the survival time is a fundamental measurement in studying certain diseases. In economics, the duration of employment periods is of great interest. In education, researchers often study the time students spend on different tasks or activities.

In studying employment histories, any one individual will generally pass through several periods of employment or unemployment, while at the same time changing their characteristics, for example the level of qualifications. From a modelling point of view, we need to model the length of time in each type of employment, relating this both to constant factors such as an individual's social origins or gender and to changing or time-dependent factors such as qualifications and age. The multilevel structure is analogous to that for repeated-measures data, with periods taking the place of occasions. Furthermore, we would generally have a further, higher level of the hierarchy since individuals, which are the level 2 units, are themselves typically clustered into workplaces, which now constitute level 3 units (formally, we can regard unemployment for this purpose as a particular workplace). In fact, the structure is even more complicated because these workplaces change from period to period and if we wish to include this level in our model we need to consider cross classifications of the units. We shall have more to say about cross classifications shortly.

There are particular problems arising when studying event duration data that are encountered when some information is 'censored' in the sense that, instead of being able to observe the actual duration, we only know that it is longer than some particular value, or in some cases less than a particular value. Chapter 9 will discuss ways of dealing with this issue for multilevel event duration models.

Discrete response data

1.6   Until now we have assumed implicitly that our response or dependent variable is continuously distributed, for example an exam score or anthropometric measure such as height. Many kinds of statistical modelling, however, deal with categorized responses, in the simplest case with proportions. Thus, we might be interested in a mortality rate, or an examination pass rate and how these vary from area to area or school to school.

In studying mortality rates in a population, it is often of great concern to try to understand the factors associated with variations from area to area or community to community. This produces a basic 2-level structure with individuals at level 1 and communities at level 2. A typical study might record deaths over a given time period together with the characteristics of the individuals concerned along with a control group and level 2 characteristics of the communities, such as their sizes or social compositions. One analysis of interest would be to see whether any of these explanatory variables could explain between-community variation. Another interest might be in studying whether mortality rate differences, say between men and women, varied from community to community.

Such models, part of the class known as generalized linear models, have been available for some time for single level data (McCullagh and Nelder, 1989), with associated software. In Chapter 7 we show how to fit multilevel models with several categorical responses and even models with mixtures of categorical and continuous responses.

Multivariate models

1.7   An interesting special case of a 2-level model is the multivariate linear (or generalized linear) model. Suppose we have taken several measurements on an individual, for example their systolic and diastolic blood pressure and their heart rate. If we wish to analyse these together as response variables we can do so by setting up a multivariate, in this case 3-variate, model with explanatory variables such as age, gender, social background, smoking exposure, etc. We can think of this as a 2-level model by considering each individual as a level 2 unit, with the three measurements constituting the level 1 units, rather as occasions did for the repeated measures model. Chapter 4 will show how this formal device for specifying a multivariate model yields considerable benefits. For example, by considering further higher levels, in this case, say, clinics, we have a simple way of specifying a multivariate multilevel model. Also, if some individuals do not have all the measurements, for example if they are randomly missing a blood pressure measurement, then this is automatically taken into account in the analysis, without the need for special procedures for handling missing data.

A particularly important application occurs where measurements are missing by design rather than at random. In certain kinds of surveys, known as rotation designs, and in certain kinds of educational assessments, known as matrix sample designs, each individual unit has only a subset of measurements made on it. For example, in large-scale testing programmes, the full range of tests may be too extensive for any one student, so that each student responds to only one combination. Such designs are viewed usefully as having a multivariate response with the full set of tests constituting the complete multivariate response vector, and every student having some tests missing. Such designs can become rather complex, especially since the students themselves are clustered into schools. By viewing the data as a single hierarchy in which the multivariate responses are level 1, we obtain an efficient and readily interpretable analysis.

The multivariate multilevel model is also the basis for ways of dealing with missing data in multilevel models and this is developed in Chapter 11.

Nonlinear models

1.8   Some kinds of data are better represented in terms of nonlinear, rather than linear, models. For example, the modelling of discrete response data is considered formally as a case of modelling nonlinear data. Many kinds of growth data are conveniently modelled in this way, especially during periods of rapid and complex growth such as early infancy and at the approach to adulthood when growth approaches an upper asymptote (Goldstein, 1979). Other examples arise when the response variable has inherent constraints. For example, biochemical activity patterns in patients may exhibit asymptotic behaviour, or cyclical patterns, both of which are difficult to model using purely linear models. Chapter 5 will introduce such models and show how to extend the linear multilevel model to this case. It will also consider cases where variances and covariances can be modelled as nonlinear functions of explanatory variables.

Measurement errors

1.9   Most measurements made in science contain some error component. This may be due to observer error as when measuring the weight of an animal, or an inherent result of being able to measure only a small sample of behaviour as in educational testing. It is well known that when variables in statistical models contain relatively large components of such error the resulting statistical inferences can be very misleading unless careful adjustments are made (Fuller, 1987). In the case of simple regression, when the explanatory or independent variable is measured with error, the usual estimate of the regression line slope is an underestimate compared with that which would result if the measurement were available without error. This is particularly important in studies of school effectiveness where the fitting of intake achievement scores is important but where such scores often have large components of measurement error.

An important case when the latter arise is where the level 2 variable is a 'compositional' variable. That is, it is a measurement aggregated from the characteristics of the level 1 units within the level 2 units. Thus, for example, the mean intake achievement and the standard deviation of the intake achievements of all the pupils in a school are compositional variables that may, and indeed sometimes do, affect the final achievements of each individual student. Likewise, in a household survey, we may consider that a measure of the average social status or the percentages of households in each social group, using all the households in the immediate community, are important explanatory variables to fit in a model. The problem arises when it is possible to collect data on only some of the level 1 units, this being the typical situation with household sample surveys. What we then have is an estimate of a compositional variable that is measured with error, in the case of household surveys typically with a very large error. In many educational studies this also occurs where only a small proportion of students within a class or school is sampled.

Chapter 10 discusses the problem of level 1 measurement error as well as the issue of measurement errors in variables measured at level 2.

Random cross classifications

1.10   Whilst the title of this book refers to multilevel, that is hierarchical, models, we have already alluded to examples where units are cross-classified as well as clustered. In geographical research, the definition of an individual's geographical area is contingent upon the context being considered. Thus, the relevant location unit for purposes of leisure may not be the same as that surrounding the environment of work or schooling. We can conceive formally of individuals belonging simultaneously to both types of unit, each of which may have an influence on a person's life.

In most schooling systems, students move from elementary to secondary or high school. We might expect that both the elementary and secondary schools attended will influence a student's achievements, behaviour and attitudes. Thus, the level 2 units are of two types, elementary school and secondary school, with each 'cell' of their cross classification containing some, or possibly no, students. In this example, a third cross classification could be the area or neighbourhood in which the student lives.

An interesting special case occurs where, for a single level 2 classification, level 1 units may belong to more than one level 2 unit. An example from sociology concerns children's and adults' friendship patterns, where an individual may belong to several groups simultaneously. The characteristics of the members of each group will influence such an individual, in relation to the individual's exposure to the group. Such multiple unit membership may be viewed formally as a multiway classification of the relevant units. Thus, for the case where an individual belongs at most to two groups we cross classify the friendship groups by themselves, with each individual belonging to one cell of the classification.

In Chapter 8 we show how to handle such random cross-classified structures as special cases of the general multilevel model. This not only allows an efficient method of modelling such structures, it also allows any complexity of mixed hierarchical and cross classified data to be handled comprehensively in the same modelling framework using the same general purpose software. For example, in epidemiological studies involving the use of trained raters or observers, a different random sample of raters may rate the status of the individuals within each level 2 unit, such as a clinic or workplace. This leads to a complex structure where, at level 1, we have a cross classification of individuals by raters, where the individuals and raters are nested within the level 2 units. Such mixtures of hierarchically structured and crossed units can be modelled within this overall framework.

Structural equation models

1.11   In many areas of the social sciences, where measurements are difficult to define precisely, an investigator might suppose that there is some underlying construct which cannot be measured directly but nevertheless can be assessed indirectly by measuring a number of relevant indicators. Structural equation modelling and, in particular, the special case of factor analysis, was developed for this purpose, typically dealing with individuals' behaviour, attitudes or mental performance. Where individuals are grouped within hierarchies, for all the same reasons discussed above, it is important to carry out such analyses in a multilevel framework. For example, we may be interested in underlying individual attitudes based upon a number of indicators. Data on such indicators may be available over time and we can postulate a model whereby the underlying attitude varies from individual to individual (level 2) and also varies randomly over time within individuals (level 1). The model can then be further elaborated by studying whether there is any systematic change over time and whether this varies across individuals. Chapter 11 discusses such models.

Levels of aggregation and ecological fallacies

1.12   When studying relationships among variables, there has often been controversy about the appropriate 'unit of analysis'. We have alluded to this already in the context of ignoring hierarchical data clustering and, as we have seen, the issue is resolved by explicit hierarchical modelling.

One of the best known early illustrations of what is often known as the ecological or aggregation fallacy was the study by Robinson (1950) of the relationship between literacy and ethnic background in the United States. When the mean literacy rates and mean proportions of Black Americans for each of nine census divisions are correlated the resulting value is 0.95, whereas the individual-level correlation, ignoring the grouping, is 0.20. Robinson was concerned to point out that aggregate-level relationships could not be used as estimates for the corresponding individual-level relationships and this point is now well understood. In Chapter 3 we shall discuss some of the statistical consequences of modelling only at the aggregate level.

Sometimes the aggregate level is the principal level of interest, but nevertheless a multilevel perspective is useful. Consider the example (Derbyshire, 1987) of predicting the proportion of children socially 'at risk' in each local administrative area for the purpose of allocating central government expenditure on social services. Survey data are available for individual children with information on risk status so that a prediction can be made using area-based variables as well as child- and household-based variables. The probability of a child being 'at risk' was estimated by the following (single level) equation

$$\text{logit}(p) = -6.3 + 5.9x_1 + 2.2x_2 + 1.5x_3$$

where x_1 is the proportion of children in the area in households with a lone parent, x_2 is the proportion of households in each area which have a density of more than 1.5 persons per room and x_3 is the proportion of households whose 'head' was born in the British 'New Commonwealth' or Pakistan. All these explanatory variables are measured at the aggregate area level and the response p is the proportion of children at risk in each area. Although we can regard this analysis as taking place entirely at the area level (with suitable weighting for the number of children in each area), there are advantages in thinking of it as a 2-level model with each child being a level 1 unit and the response variable being the binary response of whether or not the child is at risk.

First, this allows us to incorporate possibly important variables that are measured at the child level, for example whether or not each child's household is over-crowded. Including such level 1 variables may greatly improve the predictive power of the model. With the results of such a model we can then form a prediction for each area by aggregating over the known numbers of children living in over-crowded households.

Secondly, the possibility of modelling the characteristics of children or their households allows the possibility of an allocation formula that can take account of costs and benefits related to the actual composition of each area in terms of these child characteristics.

Causality

1.13 In the natural sciences, experimentation has a dominant position when making causal inferences. This is both because the units of interest can be manipulated experimentally, typically using random allocation, and because there is a widespread acceptance that the results of experiments are generalizable over space and time. The models described in this book can be applied to experimental or non-experimental data; but the final causal inferences will differ. Nevertheless, most of the examples used are from non-experimental studies in the human sciences and so a few words on causal inferences from such data may be useful.

If we wish to answer questions about a possible causal relationship between class size and educational achievement, an experimental study would need to assign different numbers of level 1 units (students) randomly to level 2 units (class teachers) and study the results over a time period of several years. This would be costly and could create ethical problems. In addition to such practical problems, any single study would be limited in time and place, and require extensive replication before results confidently could be generalized. The specific context of any study is important, for example the state of the educational system and the resources available at the time of the study. The difficulty from an experimental viewpoint is that it is practically impossible to allocate randomly with respect to all such possible confounding factors.

This is not to say that randomized experiments should never be undertaken; rather, that on their own they may have limited potential for making general statements about causality. Whether an experiment fails or succeeds in demonstrating a relationship, there will almost always be further explanations for the findings which require study. Even if an experiment appears to eliminate a possible relationship, for example demonstrating a negligible relationship between class size and attainment, it may be legitimate to query whether a relationship nevertheless exists for specific subgroups of the population.

In the pursuit of causal explanations we require some guiding underlying principles or theories. It is these which will tell us what kinds of things to measure and how to be critical of findings. For example, in studies of the relationship between perinatal mortality and maternal smoking in pregnancy (Goldstein, 1976) we can attempt to adjust for confounding factors, such as poverty, which may be responsible for influencing both smoking habits and mortality. We can also study how the relationship varies across groups and seek measures which explain such variation. We might also, in some circumstances, be able to carry out randomized experiments, assigning for example intensive health education to a randomly selected 'treatment' group and comparing mortality rates with a 'control' group.

A multilevel approach could be useful here in two different ways. First, pregnant women will be grouped hierarchically, geographically and by medical institution, and the between-area and between-institution variation may affect mortality and the relationship between mortality and smoking. Secondly, we will often be able to obtain serial measurements of smoking, so allowing the kind of repeated-measures 2-level modelling discussed earlier. This will allow us to study how changes in smoking are related to mortality, and permit a more detailed exploration of possible causal mechanisms.

Multilevel models can often be used to identify units with extreme values. For example, in school effectiveness studies an examination of school-level residual estimates (see Chapter 3) may identify those which are highly atypical, having adjusted for 'contextual' variables such as the intake characteristics of their students. These can then be selected for further examination, for example by means of intensive case studies, so forming a link between the quantitatively based multilevel analysis and a more qualitatively based investigation, which would seek to identify detailed causal processes.

A discussion of some necessary conditions for causal inference in observational studies can be found, for example, in Holland (1986) and Cochran (1983).

Finally, many of the concerns addressed by multilevel models are to do with prediction rather than causation. Thus, for example, in Chapter 6 we use a 2-level model of children's growth for the purpose of predicting adult height. In studies of school effectiveness we may be interested in understanding the causes of school differences, but we may also be concerned with predicting which school is likely to produce the best (on average) examination result for a student with given prior characteristics and achievements.

A caveat

1.14   The purpose of this book is to bring together techniques for the analysis of highly structured data, both hierarchies and cross classifications. The application of such techniques has already begun to yield new and important insights in a number of areas as the examples in the following chapters illustrate. As software becomes more widely available, the application of these techniques should become relatively straightforward, even routine.

All this is welcome, yet despite their usefulness, models for multilevel analysis cannot be a universal panacea. In some circumstances, where there is little structural complexity, such models may hardly be necessary, and traditional single level models may suffice, for both analysis and presentation. On the other hand multilevel analyses can bring extra precision to attempts to understand causality, for example by making efficient use of student achievement data in attempts to understand differences between schools. They are not, however, substitutes for well grounded substantive theories, nor do they replace the need for careful thought about the purpose of any statistical modelling. Furthermore, by introducing more complexity they can extend, but not necessarily simplify, interpretations.

Multilevel models are tools to be used with care and understanding.

2

The Basic Linear Multilevel Model and its Estimation

The 2-level model and basic notation

2.1   In this chapter, we introduce the 2-level model together with the basic notation we shall use throughout the book. We look at alternative ways of setting up and motivating the model and introduce procedures for estimating parameters, forming and testing functions of the parameters and constructing confidence intervals.

To make matters concrete, consider the following data. It is a dataset we shall also use later and it consists of 728 pupils in 48 primary (elementary) schools in inner London, part of the 'Junior School Project' (JSP). We consider two measurement occasions: the first when the pupils were in their fourth year of schooling—that is, the year they attained their eighth birthday—and three years later in their final year of primary school. Our data are in fact a subsample from a more extensive dataset which is described in detail by Mortimore *et al* (1988). We use the scores from mathematics tests administered on these two occasions together with information collected on the social background of the pupils and their gender. In this chapter, the data are used primarily to illustrate the development of basic 2-level modelling. In Chapter 3 we shall be studying more elaborate models which will enable us to handle these data more efficiently.

Figure 2.1 is a scatterplot of the 11-year-old mathematics test score by the 8-year-old test score. In this plot no distinction is made between the schools to which the pupils belong. Notice that there is a general trend, with increasing 8-year scores associated with increasing 11-year scores. Notice also the narrowing of the between pupil variation in the 11-year score with increasing 8-year score; an issue to which we shall return.

In Fig. 2.2 the scores for two particularly different schools have been selected, represented by different symbols. Two things are apparent immediately. The school represented by the circles shows a steeper 'slope' than the school represented by the filled triangles and, for most 8-year scores, the 11-year scores tend to be lower. Both these features are now addressed by formally modelling these relationships.

Consider first a simple model for one school, relating the 11-year-score to the

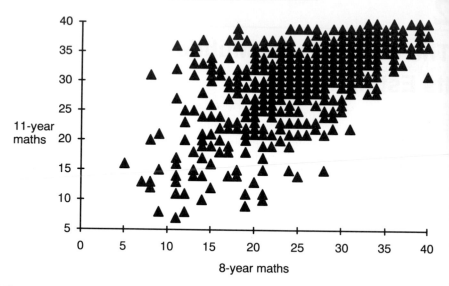

Figure 2.1   Scatterplot of 11-year by 8-year mathematics test scores. Some points represent more than one child.

8-year score. We write

$$y_i = \alpha + \beta x_i + e_i \qquad (2.1)$$

where standard interpretations can be given to the intercept (α), slope (β) and residual (e_i). We follow the normal convention of using Greek letters for the re-

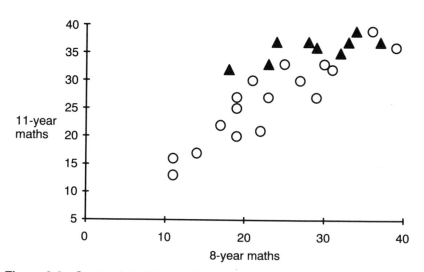

Figure 2.2   Scatterplot of 11-year by 8-year mathematics test scores for two schools.

gression coefficients and place a circumflex over any coefficient (parameter) which is a sample estimate. This is the formal model for Fig. 1.1 in the previous chapter and describes a single-level relationship. To describe simultaneously the relationships for several schools we write, for school j

$$y_{ij} = \alpha_j + \beta_j x_{ij} + e_{ij} \tag{2.2}$$

This is now the formal model for Fig. 1.2 where j refers to the level 2 unit and i to the level 1 unit.

As it stands, (2.2) is still essentially a single-level model, albeit describing a separate relationship for each school. In some situations, for example where there are few schools, and interest centres on just those schools in the sample, we may analyse (2.2) by fitting all the $2n + 1$ parameters, namely

$$(\alpha_j, \beta_j) \quad j = 1, \ldots, n \quad \sigma_e^2$$

assuming a common 'within-school' residual variance and separate lines for each school.

If we wish to focus not just on these schools, but on a wider 'population' of schools, then we need to regard the chosen schools as giving us information about the characteristics of all the schools in the population. Just as we choose random samples of individuals to provide estimates of population means etc, so a randomly chosen sample of schools can provide information about the characteristics of the population of schools. In particular, such a sample can provide estimates of the variation and covariation between schools in the slope and intercept parameters and will allow us to compare schools with different characteristics.

An important class of situations arises when we wish primarily to have information about each individual school in a sample, but where we have a large number of schools so that (2.2) would involve estimating a very large number of parameters. Furthermore, some schools may have rather small numbers of students and application of (2.2) would result in imprecise estimates. In such cases, if we regard the schools as members of a population and then use our population estimates of the mean and between-school variation, we can utilize this information to obtain more precise estimates for each individual school. This will be discussed later in the section dealing with 'residuals'.

The 2-level model

2.2 We now develop a general notation that will be used throughout this and later chapters, elaborated where necessary. We then discuss the estimation of model parameters and residuals and this is followed by illustrative examples.

To make (2.2) into a genuine 2-level model we let α_j, β_j become random variables. For consistency of notation replace α_j by β_{0j} and β_j by β_{1j} and assume that

$$\beta_{0j} = \beta_0 + u_{0j}, \quad \beta_{1j} = \beta_1 + u_{1j}$$

where u_{0j}, u_{1j} are random variables with parameters

$$\left. \begin{array}{l} E(u_{0j}) = E(u_{1j}) = 0 \\ \mathrm{var}(u_{0j}) = \sigma_{u0}^2, \quad \mathrm{var}(u_{1j}) = \sigma_{u1}^2, \quad \mathrm{cov}(u_{0j}, u_{1j}) = \sigma_{u01} \end{array} \right\} \tag{2.3}$$

We can now write (2.2) in the form

$$\left. \begin{aligned} y_{ij} &= \beta_0 + \beta_1 x_{ij} + (u_{0j} + u_{1j} x_{ij} + e_{0ij}) \\ \mathrm{var}(e_{0ij}) &= \sigma_{e0}^2 \end{aligned} \right\} \tag{2.4}$$

We shall require the extra suffix in the level 1 residual term for the models introduced in Chapter 3.

We have expressed the response variable y_{ij} as the sum of a fixed part and a random part within the brackets.

We shall also generally write the fixed part of (2.4) in the matrix form

$$E(Y) = X\beta$$

with $Y = \{y_{ij}\}$

$$E(y_{ij}) = X_{ij}\beta = (X\beta)_{ij}, \quad X = \{X_{ij}\}$$

where $\{\}$ denotes a matrix, X is the design matrix for the explanatory variables and X_{ij} is the ijth row of X. For model (2.4) we have $X = \{1 \ x_{ij}\}$. Note the alternative representation for the ith row of the fixed part of the model.

The random variables are referred to as 'residuals' and, in the case of a single-level model, the level 1 residual e_{0ij} becomes the usual linear model residual term. To make the model symmetrical so that each coefficient has an associated explanatory variable, we can define a further explanatory variable for the intercept β_0 and its associated residual, u_{0j}, namely x_{0ij}, which takes the value 1.0. For simplicity this variable may often be omitted.

The feature of (2.4) which distinguishes it from standard linear models of the regression, or analysis of variance type, is the presence of more than one residual term and this implies that special procedures are required to obtain satisfactory parameter estimates. Note that it is the structure of the random part of the model that is the key factor. In the fixed part the variables can be measured at any level, for example in the JSP data we can measure characteristics of schools or teachers. We can also include so-called 'compositional' variables such as the average 8-year mathematics test score for all pupils in each school. The presence of such variables does not alter the estimation procedure, although results will require careful interpretation.

Parameter estimation for the variance components model

2.3 Equation (2.4) requires the estimation of two fixed coefficients, β_0, β_1, and four other parameters, $\sigma_{u0}^2, \sigma_{u1}^2, \sigma_{u01}$ and σ_{e0}^2. We refer to such variances and covariances as *random parameters*. We start, however, by considering the simplest 2-level model which includes only the random parameters $\sigma_{u0}^2, \sigma_{e0}^2$. It is termed a variance components model because the variance of the response, about the fixed component, the *fixed predictor*, is

$$\mathrm{var}(y_{ij} \mid \beta_0, \beta_1, x_{ij}) = \mathrm{var}(u_0 + e_{0ij}) = \sigma_{u0}^2 + \sigma_{e0}^2$$

that is, the sum of a level 1 and a level 2 variance. For the JSP data this model implies that the total variance for each student is constant and that the covariance between two students (denoted by i_1, i_2) in the same school is given by

$$\text{cov}(u_{0j} + e_{0hj}, u_{0j} + e_{0iz}j) = \text{cov}(u_{0j}, u_{0j}) = \sigma_{u0}^2 \tag{2.5}$$

since the level 1 residuals are assumed to be independent. The correlation between two such students is therefore

$$\rho = \frac{\sigma_{u0}^2}{(\sigma_{u0}^2 + \sigma_{e0}^2)}$$

which is referred to as the 'intra-level-2-unit correlation'; in this case the intra-school correlation.[1] This correlation measures the proportion of the total variance which is between-schools. In a model with 3 levels, say with schools, classrooms and students, we will have two such correlations; the intra-school correlation measuring the proportion of variance that is between-schools and the intra-classroom correlation measuring that between classrooms.

The existence of a non-zero intra-unit correlation, resulting from the presence of more than one residual term in the model, means that traditional estimation procedures, such as 'ordinary least squares' (OLS) which are used for example in multiple regression, are inapplicable. A later section illustrates how the application of OLS techniques leads to incorrect inferences. We now look in more detail at the structure of a 2-level data set, focusing on the covariance structure typified by Fig. 2.3. The matrix in Fig. 2.3 is the (3×3) covariance matrix for the scores of three

$$\begin{pmatrix} \sigma_{u0}^2 + \sigma_{e0}^2 & \sigma_{u0}^2 & \sigma_{u0}^2 \\ \sigma_{u0}^2 & \sigma_{u0}^2 + \sigma_{e0}^2 & \sigma_{u0}^2 \\ \sigma_{u0}^2 & \sigma_{u0}^2 & \sigma_{u0}^2 + \sigma_{e0}^2 \end{pmatrix}$$

Figure 2.3   Covariance matrix of three students in a single school for a variance components model.

students in a single school, derived from the above expressions. For two schools, one with three students and one with two, the overall covariance matrix is shown in Fig 2.4. This 'block-diagonal' structure reflects the fact that the covariance between students in different schools is zero, and clearly extends to any number of level 2

$$\begin{pmatrix} A & 0 \\ 0 & B \end{pmatrix}$$

where

$$A = \begin{pmatrix} \sigma_{u0}^2 + \sigma_{e0}^2 & \sigma_{u0}^2 & \sigma_{u0}^2 \\ \sigma_{u0}^2 & \sigma_{u0}^2 + \sigma_{e0}^2 & \sigma_{u0}^2 \\ \sigma_{u0}^2 & \sigma_{u0}^2 & \sigma_{u0}^2 + \sigma_{e0}^2 \end{pmatrix}$$

$$B = \begin{pmatrix} \sigma_{u0}^2 + \sigma_{e0}^2 & \sigma_{u0}^2 \\ \sigma_{u0}^2 & \sigma_{u0}^2 + \sigma_{e0}^2 \end{pmatrix}$$

Figure 2.4   The block-diagonal covariance matrix for the response vector Y for a 2-level variance components model with two level 2 units.

$$V_2 = \begin{bmatrix} \sigma_{u0}^2 J_{(3)} + \sigma_{e0}^2 I_{(3)} & 0 \\ 0 & \sigma_{u0}^2 J_{(2)} + \sigma_{e0}^2 I_{(2)} \end{bmatrix}$$

Figure 2.5 Block-diagonal covariance matrix using general notation.

units. A more compact way of presenting this matrix, which we shall use again, is given in Fig. 2.5 where $I_{(n)}$ is the $(n \times n)$ identity matrix and $J_{(n)}$ is the $(n \times n)$ matrix of ones. The subscript 2 for V indicates a 2-level model. In single-level OLS models σ_{u0}^2 is zero and this covariance matrix then reduces to the standard form $\sigma^2 I$ where σ^2 is the (single) residual variance.

The general 2-level model including random coefficients

2.4   We can extend (2.4) in the standard way to include further fixed explanatory variables

$$y_{ij} = \beta_0 + \beta_1 x_{1ij} + \sum_{h=2}^{p} \beta_h x_{hij} + (u_{0j} + u_{1j} x_{1ij} + e_{0ij})$$

and more compactly as

$$y_{ij} = X_{ij}\beta + \sum_{h=0}^{1} u_{hj} z_{hij} + e_{0ij} z_{0ij} \tag{2.6}$$

where we use new explanatory variables for the random part of the model and write these more generally as

$$Z = \{Z_0\ Z_1\}$$
$$Z_0 = \{1\}\ \text{i.e. a vector of 1s}$$
$$Z_1 = \{x_{1ij}\}$$

The explanatory variables for the random part of the model are often a subset of those in the fixed part, as here, but this is not necessary and later we shall encounter cases where this is not so. Also, any of the explanatory variables may be measured at any of the levels; for example, we may have student characteristics at level 1 or school characteristics at level 2. Examples of both are used in the data analysis in a later section.

This model, with the coefficient of X_1 random at level 2, gives rise to the typical block structure shown in Fig. 2.6, for a level 2 block with two level 1 units. The matrix Ω_2 is the covariance matrix of the random intercept and slope at level 2. Note that we need to distinguish carefully between the covariance matrix of the responses given in Fig. 2.6 and the covariance matrix of the random coefficients. We also refer to the intercept as a random coefficient. The matrix Ω_1 is the covariance matrix for the set of level 1 random coefficients; in this case there is just a single variance term at level 1. We also write $\Omega = \{\Omega_i\}$ for the set of these covariance matrices.

$$\begin{pmatrix} A & B \\ B & C \end{pmatrix}$$

where

$$A = \sigma_{u0}^2 + 2\sigma_{u01}x_{1j} + \sigma_{u1}^2 x_{1j}^2 + \sigma_{e0}^2$$

$$B = \sigma_{u0}^2 + \sigma_{u01}(x_{1j} + x_{2j}) + \sigma_{u1}^2 x_{1j}x_{2j}$$

$$C = \sigma_{u0}^2 + 2\sigma_{u01}x_{2j} + \sigma_{u1}^2 x_{2j}^2 + \sigma_{e0}^2$$

giving

$$\begin{pmatrix} A & B \\ B & C \end{pmatrix} = X_j \Omega_2 X_j^{\mathrm{T}} + \begin{pmatrix} \Omega_1 & \\ & \Omega_1 \end{pmatrix}$$

$$X_j = \begin{pmatrix} 1 & x_{1j} \\ 1 & x_{2j} \end{pmatrix}, \quad \Omega_2 = \begin{pmatrix} \sigma_{u0}^2 & \sigma_{u01} \\ \sigma_{u01} & \sigma_{u1}^2 \end{pmatrix}, \quad \Omega_1 = \sigma_{e0}^2$$

Figure 2.6 Response covariance matrix for a level 2 unit with two level 1 units for a 2-level model with a random intercept and random regression coefficient at level 2.

We also see here the general pattern for constructing the response covariance matrix which generalizes both to higher order models and, as we shall see in Chapter 3, to complex variation at level 1. Appendix 2.1 sets out the details and describes procedures for obtaining estimates and carrying out significance tests and constructing confidence intervals for the parameters of the basic multilevel model.

Estimation for the multilevel model

2.5   We now give an overview of the Iterative Generalized Least Squares (IGLS) method which also forms the basis for many of the developments in later chapters.

We consider the simple 2-level variance components model

$$y_{ij} = \beta_0 + \beta_1 x_{ij} + u_{0j} + e_{0ij} \tag{2.7}$$

Suppose that we knew the values of the variances, and so could immediately construct the block-diagonal matrix V_2, which we will refer to simply as V. We can then apply immediately the usual Generalized Least Squares (GLS) estimation procedure to obtain the estimator for the fixed coefficients

$$\hat{\beta} = (X^{\mathrm{T}}V^{-1}X)^{-1}X^{\mathrm{T}}V^{-1}Y \tag{2.8}$$

where in this case

$$X = \begin{pmatrix} 1 & x_{11} \\ 1 & x_{21} \\ \vdots & \vdots \\ 1 & x_{n_m m} \end{pmatrix} \quad Y = \begin{pmatrix} y_{11} \\ y_{21} \\ \vdots \\ y_{n_m m} \end{pmatrix} \tag{2.9}$$

with m level 2 units and n_j level 1 units in the jth level 2 unit. When the residuals have Normal distributions (2.8) also yields maximum likelihood estimates.

Our estimation procedure is iterative. We would usually start from 'reasonable' estimates of the fixed parameters. Typically these will be those from an initial OLS fit (that is assuming $\sigma_{u0}^2 = 0$), to give the OLS estimates of the fixed coefficients $\hat{\beta}_{(0)}$. From these we form the 'raw' residuals

$$\tilde{y}_{ij} = y_{ij} - \hat{\beta}_0 - \hat{\beta}_1 x_{ij} \tag{2.10}$$

The vector of raw residuals is written

$$\tilde{Y} = \{\tilde{y}_{ij}\}$$

If we form the cross-product matrix $\tilde{Y}\tilde{Y}^T$ we see that the expected value of this is simply V. We can rearrange this cross-product matrix as a vector by stacking the columns one on top of the other, which is written as $\text{vec}(\tilde{Y}\tilde{Y}^T)$ and similarly we can construct the vector $\text{vec}(V)$. For the structure given in Fig. 2.4 these both have $3^2 + 2^2 = 13$ elements. The relationship between these vectors can be expressed as the following linear model

$$
\begin{pmatrix} \tilde{y}_{11}^2 \\ \tilde{y}_{21}\tilde{y}_{11} \\ \vdots \\ \tilde{y}_{22}^2 \end{pmatrix} = \begin{pmatrix} \sigma_{u0}^2 + \sigma_{e0}^2 \\ \sigma_{u0}^2 \\ \vdots \\ \sigma_{u0}^2 + \sigma_{e0}^2 \end{pmatrix} + R = \sigma_{u0}^2 \begin{pmatrix} 1 \\ 1 \\ \vdots \\ 1 \end{pmatrix} + \sigma_{e0}^2 \begin{pmatrix} 1 \\ 0 \\ \vdots \\ 1 \end{pmatrix} + R \tag{2.11}
$$

where R is a residual vector. The left-hand side of (2.11) is the response vector in the linear model and the right-hand side contains two explanatory variables, with coefficients σ_{u0}^2, σ_{e0}^2 which are to be estimated. The estimation involves an application of GLS using the estimated covariance matrix of $\text{vec}(\tilde{Y}\tilde{Y}^T)$, assuming Normality, namely $2(V^{-1} \otimes V^{-1})$ where $\otimes$ is the Kronecker product. The Normality assumption allows us to express this covariance matrix as a function of the random parameters. Even if the Normality assumption fails to hold, the resulting estimates are still consistent, although not fully efficient, but standard errors, estimated using the Normality assumption and, for example, confidence intervals will generally not be consistent. For certain variance component models alternative distributional assumptions have been studied, especially for discrete response models of the kind discussed in Chapter 7 (see for example Clayton and Kaldor, 1987) and maximum likelihood estimates obtained. For more general models, however, with several random coefficients, the assumption of multivariate Normality is a flexible one which allows a convenient parametrization for complex covariance structures at several levels. It is this assumption which forms the basis of the analyses in the remainder of the book.

With the estimates obtained from applying GLS to (2.11) we return to (2.8) to obtain new estimates of the fixed effects and so alternate between the random and fixed parameter estimation until the procedure converges; that is, the estimates for all the parameters do not change from one cycle to the next. Essentially the same procedure can be used for the more complicated models in the following chapters and is incorporated in the program ML3 (Prosser *et al*, 1991) and its more general successor MLn (Rasbash *et al*, 1995).

The maximum likelihood procedure produces biased estimates of the random parameters because it takes no account of the sampling variation of the fixed parameters. This may be important in small samples, and we can produce unbiased estimates by using a modification known as restricted maximum likelihood (REML). The IGLS algorithm is readily modified to produce these restricted estimates (RIGLS) (Goldstein, 1989a).

Other estimation procedures

2.6   Longford (1987) developed a procedure based upon a 'Fisher scoring' algorithm and Raudenbush (1994) showed that it is formally equivalent to IGLS. A program VARCL (Longford, 1987) uses this algorithm and also incorporates certain extensions, for example to handle discrete response data (see Chapter 7). A variation on IGLS is Expected Generalized Least Squares (EGLS). This focuses interest on the fixed part parameters and uses the estimate of V obtained after the first iteration merely to obtain a consistent estimator of the fixed part coefficients without further iterations. A variant of this separates the level 1 variance from V as a parameter to be estimated iteratively along with the fixed part coefficients.

A rather different approach is to view (2.2), and more general extensions, as a Bayesian linear model (Lindley and Smith, 1972) where the β_j are assumed to be exchangeable and to have a prior distribution with variance σ_{u0}^2. The full Bayes estimation then requires a prior distribution for the random parameters also, in this case the level 1 and level 2 variances. An alternative to the full Bayes estimation, known as 'empirical Bayes', ignores the prior distributions of the random parameters, treating them as known for purposes of inference. When Normality is assumed, these estimates are the same as IGLS or RIGLS. Bryk and Raudenbush (1992) describe the use of an EM algorithm to provide such estimates, and the program HLM (see Chapter 11) uses this algorithm.

Another approach which parallels all of these is that of Generalized Estimating Equations (GEE) introduced by Liang and Zeger (1986). The principal difference is that GEE obtains the estimate of V using simple regression or 'moment' procedures based upon functions of the actual calculated raw residuals. It is concerned principally with modelling the fixed coefficients rather than exploring the structure of the random component of the model. While the resulting coefficient estimates are consistent they are not fully efficient. In some circumstances, however, GEE coefficient estimates may be preferable, since they will usually be quicker to obtain and they make weaker assumptions about the structure of V. The GEE procedure can be extended to handle most of the models dealt with in later chapters.

More recently, the full Bayesian treatment has become computationally feasible with the development of 'Markov Chain Monte Carlo' (MCMC) methods, especially Gibbs Sampling (Zeger and Karim, 1991). This has the advantage, in small samples, that it takes account of the uncertainty associated with the estimates of the random parameters and can provide exact measures of uncertainty. The maximum likelihood methods tend to overestimate precision because they ignore this uncertainty. In small samples this will be important especially when obtaining 'posterior' estimates for residuals, which we deal with later in the chapter. In Chapter 3 we present an alternative 'bootstrap' procedure for taking account of this uncertainty.

Appendix 2.4 provides details of Gibbs Sampling and Appendix 2.3 of empirical Bayes estimates.

We shall have more to say about computational issues in Chapter 11.

Residuals

2.7   In a single level model such as (2.1) the usual estimate of the single residual term e_i is just $\tilde{y}_i$, the raw residual. In a multilevel model, however, we shall generally have several residuals at different levels. We consider estimating the individual residuals along the following lines.

Given the parameter estimates, consider predicting a specific residual, say u_{0j} in a 2-level variance components model. Specifically, we require for each level 2 unit

$$\hat{u}_{0j} = E(u_{0j} \mid Y, \hat{\beta}, \hat{\Omega}) \tag{2.12}$$

We shall refer to these as estimated or predicted residuals or, using Bayesian terminology, as posterior residual estimates. If we ignore the sampling variation attached to the parameter estimates in (2.12) we have

$$\left.\begin{aligned}
\text{cov}(\tilde{y}_{ij}, u_{0j}) &= \text{var}(u_{0j}) = \sigma_{u0}^2 \\
\text{cov}(\tilde{y}_{ij}, e_{0ij}) &= \sigma_{e0}^2 \\
\text{var}(\tilde{y}_{ij}) &= \sigma_{u0}^2 + \sigma_{e0}^2
\end{aligned}\right\} \tag{2.13}$$

We regard (2.12) as a linear regression of u_{0j} on the set of $\{\tilde{y}_{ij}\}$ for the jth level 2 unit and (2.13) defines the quantities required to estimate the regression coefficients and hence $\hat{u}_{0j}$. Details are given in Appendix 2.2. For the variance components model we obtain

$$\left.\begin{aligned}
\hat{u}_{0j} &= \frac{n_j \sigma_u^2}{(n_j \sigma_u^2 + \sigma_{e0}^2)} \tilde{y}_j \\
\tilde{e}_{0ij} &= \tilde{y}_{ij} - \hat{u}_{0j} \\
\tilde{y}_j &= (\sum_i \tilde{y}_{ij}) / n_j
\end{aligned}\right\} \tag{2.14}$$

where n_j is the number of level 1 units in the jth level 2 unit. The residual estimates are not, unconditionally, unbiased but they are consistent. The factor multiplying the mean ($\tilde{y}_j$) of the raw residuals for the jth unit is often referred to as a 'shrinkage factor' since it is always less than or equal to one. As n_j increases this factor tends to one, and as the number of level 1 units in a level 2 unit decreases the 'shrinkage estimator' of u_{0j} becomes closer to zero. In many applications the higher level residuals are of interest in their own right and the increased shrinkage for a small level 2 unit can be regarded as expressing the relative lack of information in the unit so that the best estimate places the predicted residual close to the overall population value as given by the fixed part.

These residuals therefore can have two roles. Their basic interpretation is as random variables with a distribution whose parameter values tell us about the variation among the level 2 units, and which provide efficient estimates for the fixed coefficients. A second interpretation is as individual estimates for each level 2 unit where we use the assumption that they belong to a population of units to predict their values. In particular, for units which have only a few level 1 units, we can obtain more precise estimates than if we were to ignore the population membership assumption and use only the information from those units. This becomes especially important for estimates of residuals for random coefficients, where in the extreme case of only one level 1 unit in a level 2 unit we lack information to form an independent estimate. In Chapter 6 we shall illustrate this when we consider predictions based upon repeated measures growth models.

As in single level models we can use the estimated residuals to help check on the assumptions of the model. The two particular assumptions that can be studied readily are the assumption of Normality and that the variances in the model are constant. Because the variances of the residual estimates depend, in general, on the values of the fixed coefficients it is common to standardize the residuals by dividing by the appropriate standard errors. The formulae for these are given in Appendix 2.2, where we refer to them as 'diagnostic' or 'unconditional' standard errors.

When the residuals at higher levels are of interest in their own right, we need to be able to provide interval estimates and significance tests as well as point estimates for them or functions of them. For these purposes we require estimates of the standard errors of the estimated residuals, where the sample estimate is viewed as a random realization from repeated sampling of the same higher level units whose unknown true values are of interest. The formulae for these 'conditional' or 'comparative' standard errors are also given in Appendix 2.2.

The level 1 residuals are generally not of interest in their own right but are used rather for model checking, having first been standardized using the diagnostic standard errors.

The adequacy of Ordinary Least Squares estimates

2.8   In Appendix 2.1 we give the formulae for estimating the true standard errors for OLS estimates when a multilevel model applies. When the intra-unit correlations are small we can expect reasonably good agreement between the multilevel estimates and the simpler OLS ones. While it is difficult to give general guidelines about when OLS is an adequate alternative we can readily derive an explicit formula for the balanced 2-level variance components model using a simple regression equation with an intercept and a single explanatory variable

$$y_{ij} = \beta_0 + \beta_1 x_{ij} + u_j + e_{ij}$$

Write ρ_y, ρ_x for the intra-unit correlations for Y, X respectively and n for the number of level 1 units in the jth level 2 unit. To obtain an estimate of the correct standard error for the estimate of β_1 we multiply the usual OLS estimate of the standard error by the quantity

$$\left\{ 1 + \rho_y \rho_x [n - 1] \right\}^{\frac{1}{2}}$$

where m is the number of level 2 units. Thus, if there is exactly one level 1 unit per level 2 unit or either of the intra-unit correlations are zero, this expression is equal to 1.0 and the usual expression is correct. As n increases so the OLS estimator increasingly underestimates the true standard error. Thus, with $\rho_y = \rho_x = 0.20$ and 76 level 1 units per level 2 unit, the true standard error is, on average, twice the OLS estimate. Hence, confidence intervals based on the OLS estimate will be too short and significance tests will too often reject the null hypothesis. By designing a study where the n is small we may be able to rely on OLS procedures to give adequate estimates for the fixed coefficients, but this does not then allow us to study any multilevel structures.

A 2-level example using longitudinal educational achievement data

2.9   We shall fit the simple 2-level variance components model (2.7) to the JSP data with the 11-year maths score as response and a single explanatory variable, the 8-year maths score, in addition to the constant term, equal to 1 and defining the intercept. The parameter values are displayed in Table 2.1 with the Ordinary Least Squares estimates given for comparison.

Table 2.1 Variance components model applied to JSP data

Parameter	Estimate (s.e.)	OLS Estimate (s.e.)
Fixed:		
Constant	13.9	13.8
8-year score	0.65 (0.025)	0.65 (0.026)
Random:		
σ_{u0}^2 (between schools)	3.19 (1.0)	
σ_{e0}^2 (between students)	19.8 (1.1)	23.3 (1.2)
Intra-school correlation	0.14	

Comparing the OLS with the multilevel estimates we see that the fixed coefficients are similar, but that there is an intra-school correlation of 0.14. The estimate of the standard error of the between school variance is less than a third of the variance estimate, suggesting a value highly significantly different from zero. This comparison, however, should be treated cautiously, since the variance estimate does not have a Normal distribution and the standard error is only estimated, although the size of the sample here will make the latter caveat less important. It is generally preferable to carry out a likelihood ratio test by estimating the 'deviance' for the current model and the model omitting the level 2 variance (see McCullagh and Nelder, 1989). The next section will deal more generally with inference procedures. The deviances are, respectively, 4294.2 and 4357.3 with a difference of 63.1, which is referred to tables of the chi-squared distribution with one degree of freedom, and is highly significant. Note that if we use the standard error estimate given in Table 2.1 to judge significance we obtain the corresponding value of $(3.19/1.0)^2 = 10.2$ which is very much smaller than the likelihood ratio test statistic.

Table 2.2 Variance components model applied to JSP data with gender and social class

Parameter	Estimate (s.e.)	Estimate (s.e.)
Fixed:		
Constant	14.9	32.9
8-year score	0.64 (0.025)	
Gender (boys–girls)	−0.36 (0.34)	−0.39 (0.47)
Social Class (Non Man.–Manual)	0.72 (0.39)	2.93 (0.51)
Random:		
σ_{u0}^2 (between schools)	3.21 (1.0)	4.52 (1.5)
σ_{e0}^2 (between students)	19.6 (1.1)	37.2 (2.0)
Intra-school correlation	0.14	0.11

We elaborate the model first by adding two more explanatory variables, gender and social class. The results are set out in the first column of Table 2.2.

The random parameter estimates are hardly changed, nor is the coefficient of the 8-year maths score. The gender difference is very small and in favour of the girls, but is far from the conventional 5% significance level. The social class difference favours the children of non-manual parents. When we are judging the fixed effects, a simple comparison of the estimate with its standard error is usually adequate. Because the model adjusts for the earlier maths score we can interpret the social class and gender differences in terms of the relative *progress* of girls versus boys or non-manual versus manual children. The second column in Table 2.2 shows the effects when the 8-year maths score is removed from the model and the interpretation is now in terms of the actual differences found at 11 years. Note that the level 1 and level 2 variances are increased, reflecting the importance of the earlier score as a predictor, and the intra-school correlation is slightly reduced. The social class difference is much larger, suggesting that most of the difference is that existing at 8 years with a somewhat greater progress made between 8 and 11 years by those in the non-manual social group. The gender difference remains small.

The 8-year score has been used as it stands, without centring it in any way. This is acceptable in the present case, although the strict interpretation of the intercept is the predicted score at an 8-year score of zero, which is outside the range of the observed values. If we were to measure the 8-year score from its mean, the intercept would be interpreted as the predicted value at the mean 8-year score. When we introduce random coefficients in Chapter 3 we shall see that this becomes an important consideration.

Checking model assumptions

2.9.1   We now check some assumptions of the model by looking at the residuals. Figure 2.7 is a plot of the standardized level 1 residuals against the fixed part predicted value and Fig. 2.8 is a plot of these residuals against their equivalent Normal scores. Figure 2.9 is the equivalent Normal score plot for the level 2 residuals. Figure 2.7 shows the same pattern as Fig. 2.1 of a decreasing variance with increasing 8-year score, so that the assumption of a constant level 1 variance is clearly un-

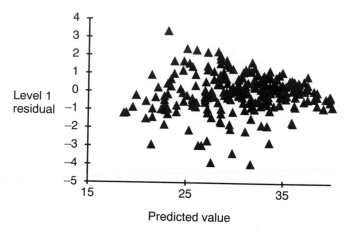

Figure 2.7   Standardized level 1 residuals by predicted values for Table 2.2.

tenable. In Chapter 3 we shall be looking at ways to deal with this. The Normal score plots, on the other hand, are fairly straight, suggesting that the Normal distribution assumption is reasonable for both level 1 and level 2.

Checking for influential units

2.9.2   Inspection of Fig. 2.9 shows one school, identified as number 38, with the largest standardized residual and unstandardized value of 3.5 compared with 2.9 for the next largest. It is often useful to study the effect of omitting one or more units from an analysis to see what difference this makes to the parameter estimates. Efficient techniques, known as 'influence analysis', for deciding which units to

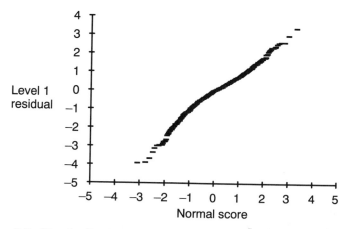

Figure 2.8   Standardized level 1 residuals by Normal equivalent scores for Table 2.2.

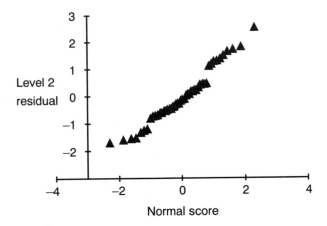

Figure 2.9 Standardized level 2 residuals by Normal equivalent scores for Table 2.2.

treat in this way are well developed for single level models (Cook and Weisberg, 1982). Efficient techniques are not yet available for multilevel models, but the effect of omitting specific units can be studied. We illustrate this for school 38. Table 2.3 shows the parameter estimates associated with two different procedures.

In analysis A, school 38 is simply omitted. The principal effect is to reduce the level 2 variance by about 14%, with little effect on the other parameters. In analysis B we have retained all the data in the analysis, but removed school 38 from the level 2 variation by fitting a separate constant in the fixed part of the model. For the explanatory variable defining the level 2 variance we fit Z_0^* rather than Z_0, where

$$Z_0^* = \begin{cases} 0 \text{ if school 38} \\ 1 \text{ otherwise} \end{cases}$$

Table 2.3 As Table 2.2. Analysis A omitting school 38. Analysis B fitting a constant for school 38

Parameter	Estimate (s.e.) A	Estimate (s.e.) B
Fixed:		
Constant	14.5	14.7
8-year score	0.65 (0.026)	0.64 (0.025)
Gender (boys–girls)	−0.40 (0.34)	−0.37 (0.34)
Social Class (Non Man.–Manual)	0.74 (0.39)	0.72 (0.38)
School 38		6.1 (1.5)
Random:		
σ_{u0}^2 (between schools)	2.74 (0.9)	2.75 (0.9)
σ_{e0}^2 (between students)	19.6 (1.1)	19.6 (1.1)
Intra-school correlation	0.12	0.12

and the constant fitted in the fixed part is simply $1 - Z_0^*$. The relatively small number of students, nine, in school 38 accounts for the fact that its shrunken residual mean of 3.5 is considerably less than the directly fitted mean of 6.1. Although it makes little difference to the parameter estimates in this example, in general it seems preferable to fit separate parameters for influential units and retain as much data as possible in the analysis.

Higher level explanatory variables and compositional effects

2.10   We have already mentioned that, from the point of view of estimating parameters, the explanatory variables can be defined or measured at any level. For substantive interpretations, however, explanatory variables measured at levels 2 or above often have particular interpretations. We illustrate some of these by using the JSP dataset and forming the explanatory variable which is the mean 8-year-old maths score. This is often known as a 'compositional' variable since it measures an aspect of the composition of the school to which the individual student belongs. We are interested in whether the average 8-year score has an effect on the 11-year score, after having adjusted for the student's own 8-year score. For this analysis all the 8-year scores are measured about the sample mean value of 25.98, see Table 2.4.

Analysis A adds the average school 8-year score. Its coefficient is very small and not significant. Analysis B uses the school centred 8-year score. This is often advocated on the grounds that it is the difference between a student's score and the average score for that student's school which is likely to be the most relevant predictor of later achievement. Bryk and Raudenbush (1992, Chapter 5) gave a

Table 2.4 Variance components model for JSP data with mean 8-year score measured about sample mean and centring about school mean

Parameter	Estimate (s.e.) A	Estimate (s.e.) B	Estimate (s.e.) C
Fixed:			
Constant	31.5	31.5	31.7
8-year score	0.64 (0.025)		0.63 (0.025)
8-year score centred on school mean		0.64 (0.026)	
Gender (boys–girls)	−0.36 (0.34)	−0.36 (0.34)	−0.37 (0.34)
Social Class (Non Man.–Manual)	0.72 (0.38)	0.72 (0.31)	0.79 (0.31)
School mean 8-year score	−0.01 (0.13)	0.63 (0.12)	−0.03 (0.12)
8-year score × school mean 8-year score			−0.02 (0.01)
Random:			
σ_{u0}^2 (between schools)	3.21 (1.0)	3.21 (1.0)	3.13 (1.0)
σ_{e0}^2 (between students)	19.6 (1.1)	19.6 (1.0)	19.5 (1.1)
Intra-school correlation	0.14	0.14	0.14

detailed discussion of this issue for models where the compositional variable, as here, is a mean computed for all the students in the school, or more generally all the level 1 units in the relevant level 2 unit. Analyses A and B are, of course, formally equivalent and analysis A indicates directly that a simpler model omitting the school mean score is adequate. It is analysis C, as discussed below, which introduces a more complex model.

In fact, the mean score for students in a school is only one particular summary statistic describing the composition of the students. Another summary would be the spread of scores, measured for example by their standard deviation. We can also consider measures such as the proportions of high or low scoring students and, in general, any set of such measures. When using the average score we can also consider using the median or modal score rather than the mean. With any of these other measures we may wish to retain the deviation from the school mean as an explanatory variable, and we could even consider introducing a more complex function of this, for example by adding higher order terms. There is here a fruitful area for further study.

Analysis C looks at the possibility of an interaction between student score and school mean and we do find a significant effect which we can interpret as follows. The higher the school mean 8-year score the lower the coefficient of the student's 8-year score. One implication of this is that for two relatively low scoring students at 8 years, the one in the school with a higher average is predicted to do better at 11 years. To study this further we now need to introduce a model with random coefficients where we explicitly allow each school's coefficient to vary randomly at level 2, as in Equation (2.6): see Table 2.5.

The addition of the 8-year score coefficient as a random variable at level 2 somewhat increases the social class difference and somewhat decreases the gender difference, but within their standard errors. The level 1 variance is reduced and we

Table 2.5 Random coefficient model for JSP data

Parameter	Estimate (s.e.)
Fixed:	
Constant	31.7
8-year score	0.62 (0.036)
Gender (boys–girls)	−0.25 (0.32)
Social Class (Non Man.–Manual)	0.96 (0.36)
School mean 8-year score	−0.04 (0.13)
8-year score × school mean 8-year score	−0.02 (0.01)
Random:	
Level 2	
σ_{u0}^2 (Intercept)	3.67 (1.03)
σ_{u01} (covariance)	−0.34 (0.09)
σ_{u1}^2 (8-year score)	0.03 (0.01)
Level 1	
σ_{e0}^2	17.8 (1.0)

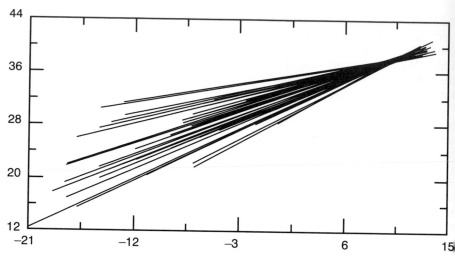

Figure 2.10   Plot of predicted 11-year score by 8-year score for JSP schools.

have significant 'slope' variation at level 2; the likelihood ratio test criterion is 52.4, which is referred to chi-squared tables with 2 degrees of freedom and is highly significant. If we calculate the correlation between the intercept and slope at level 2 we obtain a value of -1.03! This sometimes happens as a result of sampling variation and implies that the population correlation is very high. We shall see in Chapter 3 that we can constrain this correlation to be exactly -1.0 and thus admissible. Alternatively, by suitably elaborating the model or by carrying out certain transformations we can avoid this problem. For now, however, in order to illustrate what this means in the present data we can compute residuals for each school, for the slope and intercept. With these estimates we can then predict the 11-year score for any set of values of the explanatory variables. Figure 2.10 shows the predicted values for manual girls by 8-year score. The predicted lines for the high scores at 8 years are very close together and separate as the 8-year score decreases. The slope residual is almost uncorrelated (-0.02) with the mean 8-year score and the compositional coefficient of the mean 8-year score is little changed. We can add, therefore, to the previous compositional effect, the statement that some schools are differentially 'effective' for pupils with low 8-year scores, with little difference for high 8-year scores. In Chapter 3 we shall continue to analyse this dataset and show how further elaboration of the variance structure of the model leads to certain simplifications of interpretation.

Hypothesis testing and confidence intervals

2.11   In this section we deal with large sample procedures for constructing interval estimates for parameters, or linear functions of parameters, and for hypothesis testing. Hypothesis tests are used sparingly throughout this book, since the usual form of a null hypothesis, that a parameter value or a function of parameter values is zero, is usually implausible and also relatively uninteresting. Moreover, with large enough samples a null hypothesis will almost certainly be rejected. The exception to this is where we are interested in whether a difference is positive or negative, and this is discussed in the section on residuals below. Confidence intervals emphasize the uncertainty surrounding the parameter estimates and the importance of their substantive significance.

Fixed parameters

2.11.1   In the analyses of 2.11 we presented parameter estimates for the fixed part parameters together with their standard errors. These are adequate for hypothesis testing, or confidence interval construction, separately for each parameter. In many cases, however, we are interested in combinations of parameters. For hypothesis testing, this most often arises for grouped or categorized explanatory variables where n group effects are defined in terms of $n-1$ dummy variable contrasts and we wish simultaneously to test whether these contrasts are zero. In the case of the analysis in Table 2.2 we may be interested in the hypothesis that the gender and social class effects, taken jointly, are zero. We may also be interested in providing a pair of confidence intervals for the parameter estimates. We proceed as follows.

Define an $(r \times p)$ contrast matrix C. This is used to form linearly independent functions of the p fixed parameters in the model of the form $f = C\beta$, so that each row of C defines a particular linear function. Parameters which are not involved have the corresponding elements set to zero. Suppose we wish to test the hypothesis in Table 2.2 that the gender and social class coefficients are jointly zero. We define

$$C = \begin{pmatrix} 0 & 0 & 1 & 0 \\ 0 & 0 & 0 & 1 \end{pmatrix}, \quad f = \begin{pmatrix} \beta_2 \\ \beta_3 \end{pmatrix}$$

and the general null hypothesis is

$$H_0 : f = k, \quad k = \{0\} \text{ here}$$

We form

$$R = (\hat{f} - k)^{\mathrm{T}} [C(X^{\mathrm{T}} \hat{V}^{-1} X)^{-1} C^{\mathrm{T}}]^{-1} (\hat{f} - k)$$

$$\hat{f} = C\hat{\beta}$$

(2.15)

If the null hypothesis is true this is distributed as approximately χ^2 with r degrees of freedom. Note that the term $(X^{\mathrm{T}} \hat{V}^{-1} X)^{-1}$ is the estimated covariance matrix of the fixed coefficients.

If we find a statistically significant result we may wish to explore which particular linear combinations of the coefficients involved are significantly different from zero. The common instance of this is where we find that n groups differ and we wish to carry out all possible pairwise comparisons. A simultaneous comparisons procedure, which maintains the overall Type I error at the specified level, involves carrying out the above procedure with either a subset of the rows of C or a set of (less than r) linearly independent contrasts. The value of R obtained is then judged against the critical values of the chi-squared distribution with r degrees of freedom.

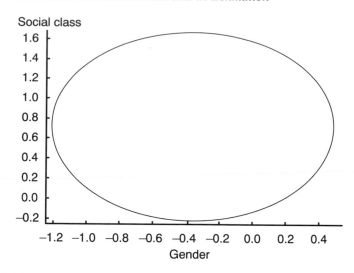

Figure 2.11   95% confidence region for coefficients of social class and gender.

We can also obtain an $\alpha\%$ confidence region for the parameters by setting $\hat{R}$ equal to the $\alpha\%$ tail region of the χ^2 distribution with r degrees of freedom in the expression

$$\hat{R} = (f - \hat{f})^{\mathrm{T}}[C(X^{\mathrm{T}}\hat{V}^{-1}X)^{-1}C^{\mathrm{T}}]^{-1}(f - \hat{f})$$

This yields a quadratic function of the estimated coefficients, giving an r-dimensional ellipsoidal region. For Table 2.2 we obtain the following results.

The null hypothesis test gives a value for chi squared on 2 degrees of freedom of 4.51 with a corresponding P-value of 0.10. The 95% confidence region is the ellipse

$$8.3(\beta_1 + 0.36)^2 + 0.22(\beta_1 + 0.36)(\beta_2 - 0.72) + 6.7(\beta_2 - 0.72)^2 = 5.99$$

where the subscripts (1,2) refer to gender and social class respectively and 5.99 is the 5% point of the χ_2^2 distribution. Figure 2.11 displays this region.

In some situations we may be interested in separate confidence intervals for all possible linear functions involving a subset of q parameters or q linearly independent functions of the parameters, while maintaining a fixed probability that all the intervals include the population value of these functions of the parameters. As before, this may arise when we have an explanatory variable with several categories and we are interested in intervals for sets of contrasts. For a $(1-\alpha)\%$ interval, write C_i for the ith row of C, and then a simultaneous $(1-\alpha)\%$ interval for $C_i\beta$, for all C_i, is given by

$$(C_i\hat{\beta} - d_i, C_i\hat{\beta} + d_i)$$

where

$$d_i = [C_i(X^{\mathrm{T}}\hat{V}^{-1}X)^{-1}C_i^{\mathrm{T}}\chi^2_{q,(\alpha)}]^{0.5}$$

where $\chi^2_{q,(\alpha)}$ is the $\alpha\%$ point of the χ^2_q distribution.

For model A of Table 2.2 we obtain the following 95% intervals for the coefficients of gender and social class, first the separate intervals then the simultaneous ones which are some 25% wider.

$$\begin{pmatrix} -0.36 \pm 0.66 \\ 0.72 \pm 0.76 \end{pmatrix}, \quad \begin{pmatrix} -0.36 \pm 0.83 \\ 0.72 \pm 0.94 \end{pmatrix}$$

We can also use the likelihood ratio test criterion for testing hypotheses about the fixed parameters, although generally the results will be similar. The difference arises because the random parameter estimates used in (2.15) are those obtained for the full model rather than those under the null hypothesis assumption, although this modification can easily be made. For example, the likelihood ratio test for gender and social class yields a value of 5.5 compared with the above value of 4.5. We shall discuss the likelihood ratio test in the next section dealing with the random parameters.

Random parameters

2.11.2   In very large samples it is possible to use the same procedures for hypothesis testing and confidence intervals as for the fixed parameters. Generally, however, procedures based upon the likelihood statistic are preferable. To test a null hypothesis H_0 against an alternative H_1 involving the fitting of additional parameters, we form the log-likelihood ratio or deviance statistic

$$D_{01} = -2 \log_e (\lambda_0 / \lambda_1) \qquad (2.16)$$

where λ_0, λ_1 are the likelihoods for the null and alternative hypotheses and this is referred to tables of the chi-squared distribution with degrees of freedom equal to the difference (q) in the number of parameters fitted under the two models. We have already quoted this statistic for testing the level 2 variance in Table 2.1 where the value of 63.1 compares with the statistic formed by taking the variance estimate and dividing by its standard error and then squaring the result to give a value of 11.0.

We can also use (2.16) as the basis for constructing a $(1 - \alpha)\%$ confidence region for the additional parameters. If D_{01} is set to the value of the $\alpha\%$ point of the chi-squared distribution with q degrees of freedom, then a region is constructed to satisfy (2.16), using a suitable search procedure. This is a computationally intensive task, however, since all the parameter estimates are recomputed for each search point.

If we carry out these calculations for the level 2 variance in Table 2.1 we obtain a 95% confidence interval of (1.78, 5.65). Likewise we can obtain an interval for the intra-school correlation by searching in two dimensions and computing the value at each search point. This gives a 95% confidence interval of (0.09, 0.22). A review of some approximate procedures is given by Burdick and Graybill (1988).

An alternative is to use the 'profile likelihood' (McCullagh and Nelder, 1989). In this case the likelihood is computed for a suitable region containing values of the random parameters of interest, for fixed values of the remaining random parameters. For the level 2 variance of Table 2.1 this gives a 95% confidence interval of (1.77, 5.69) which is very close to the full likelihood interval.

In Chapter 3 we shall see how bootstrap simulations can provide interval estimates.

Residuals

2.11.3   In our JSP variance components analysis we estimated level 2 residuals, one for each school. In studies of school effectiveness, one requirement is sometimes to try to identify schools with residuals which are substantially different. From a significance testing standpoint, we will often be interested in the null hypothesis that school A has a smaller residual than school B against the alternative that the residual for school A is larger than that for school B (ignoring the vanishingly small probability that they are equal). In the case when a standard significance test accepts the alternative hypothesis (at a chosen level) of some difference against the null hypothesis of no difference, this is equivalent to accepting one of the alternatives (A > B, A < B) at the same level of significance and we shall use this interpretation.

Where we can identify *a priori* two schools then it is straightforward, using the results of Appendix 2.1, to construct a confidence interval for their difference or carry out a significance test. Often, however, the results are made available to a number of individuals, each of whom is interested in comparing their own schools of interest. This may occur, for example, where policy makers wish to select a few schools within a small geographical area for comparison, out of a much larger comparative study. In the following discussion, we suppose that individuals wish to compare only pairs of schools, although the procedure can be extended to multiple comparisons of three or more residuals. Further details are given by Goldstein and Healy (1994).

Consider the JSP data where we have 48 estimated residuals together with their comparative standard errors. Since the sample size is fairly large, we can also assume that these estimates are uncorrelated.

First, we order the residuals from smallest to largest. We construct an interval about each residual so that the criterion for judging statistical significance at the $(1 - \alpha)\%$ level for any pair of residuals is whether their confidence intervals overlap. For example, if we consider a pair of residuals with a common standard error (s.e.) , and assuming Normality, the confidence interval width for judging a difference significant at the 5% level is given by ± 1.39(s.e.).

The general procedure defines a set of confidence intervals for each residual i as

$$\hat{u}_i \pm c(\text{s.e.})_i \tag{2.17}$$

For each possible pair of intervals (2.17), there is a significance level associated with the overlap criterion, and the value c is determined so that the average over all possible pairs is $(1 - \alpha)\%$. A search procedure can be devised to determine c. When the ratios of the standard errors do not vary appreciably, say by not more than 2:1, the value 1.4 can be used for c. As this ratio increases so does the value of c. In the present case, all but two of these ratios are greater than 2 and we have used the common value of 1.4.

The results are presented in Fig. 2.12. As is clear, apart from some of the extreme intervals, each interval overlaps with most of the other intervals. If we wished the basic comparison to take place among triplets of schools, with simultaneous confidence intervals, then using the results of section 2.11.1 we replace the

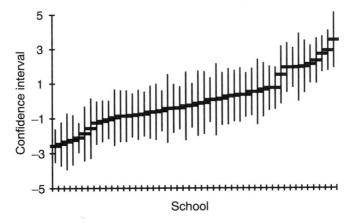

Figure 2.12 Simultaneous confidence intervals for JSP school residuals.

Normal upper 2.5% value of 1.96 by $\sqrt{\chi^2_{2,(0.05)}} = 2.45$. This will give a similar display but with intervals 25% wider. In reality, the complete set of schools will typically be compared in overlapping subsets of different sizes, and a value for c can be determined by averaging over all such possibilities.

Presentations such as that in 2.12 are useful for conveying the inherent uncertainty associated with estimates for individual level 2 (or higher) units, where the number of level 1 units per higher level unit is not large. This uncertainty, in turn, places inherent limitations upon such comparisons.

Note

1. In the sample survey literature and elsewhere, such as in genetics, the term 'intra-class correlation' is used, but this clearly is confusing in the present context.

APPENDIX 2.1

The general structure and estimation for a multilevel model

We illustrate the general structure using a 2-level model. We have

$$
\left.
\begin{aligned}
& Y = X\beta + E \\
& Y = \{y_{ij}\}, \quad X = \{X_{ij}\}, \quad X_{ij} = \{x_{0ij}, x_{1ij}, \ldots, x_{pij}\} \\
& E = E_1 + E_2 = \{e_{ij}\}, \quad e_{ij} = e_{ij}^{(1)} + e_j^{(2)} \\
& e_{ij}^{(1)} = \sum_{h=0}^{q_1} z_{hij}^{(1)} e_{hij}^{(1)}, \quad e_j^{(2)} = \sum_{h=0}^{q_2} z_{hij}^{(2)} e_{hj}^{(2)}
\end{aligned}
\right\}
\qquad (2.1.1)
$$

We will also write simply

$$
e_{ij}^{(1)} = e_{ij}, \quad e_j^{(2)} = u_j
$$

$$
Y = X\beta + Z^{(2)}u + Z^{(1)}e
$$

The residual matrices E_1, E_2 have expectation zero with

$$
\begin{aligned}
& E(E_1 E_1^\mathrm{T}) = V_{2(1)}, \quad E(E_2 E_2^\mathrm{T}) = V_{2(2)} \\
& E(E_1 E_2^\mathrm{T}) = 0, \qquad V_2 = V_{2(1)} + V_{2(2)}
\end{aligned}
\qquad (2.1.2)
$$

In the standard model the level 1 residuals are assumed independent across level 1 units, so that $V_{2(1)}$ is diagonal with ijth element

$$
\mathrm{var}(e_{ij}) = \sigma_{eij}^2 = z_{ij}^{(1)^\mathrm{T}} \Omega_e z_{ij}^{(1)}, \quad \Omega_e = \mathrm{cov}(e_h^{(1)})
$$

The level 2 residuals are assumed independent across level 2 units and $V_{2(2)}$ is block-diagonal with jth block

$$
V_{2(2)j} = z_j^{(2)^\mathrm{T}} \Omega_u z_j^{(2)}, \quad \Omega_u = \mathrm{cov}(e_h^{(2)})
$$

The jth block of V_2 is therefore given by

$$
V_{2j} = \oplus_i \sigma_{eij}^2 + V_{2(2)j}
\qquad (2.1.3)
$$

where $\oplus$ is the direct sum operator.

For some of the models dealt with in later chapters, such as the time series models of Chapter 6, the requirement of independence among the residuals for the level 1 units is relaxed. In this case the first term on the right-hand side of (2.1.3) is replaced by the particular structure of $V_{2(1)}$.

For known V_2 and omitting the subscript for convenience, the generalized least squares estimate of the fixed coefficients is

$$\hat{\beta} = (X^{\mathrm{T}}V^{-1}X)^{-1}X^{\mathrm{T}}V^{-1}Y \qquad (2.1.4)$$

with covariance matrix

$$(X^{\mathrm{T}}V^{-1}X)^{-1}$$

For known β we form

$$Y^* = \tilde{Y}\tilde{Y}^{\mathrm{T}}, \quad \tilde{Y} = Y - X\beta = E_1 + E_2 \qquad (2.1.5)$$

and we have $E(Y^*) = V$. We now write

$$Y^{**} = \mathrm{vec}(Y^*)$$

where vec is the vector operator stacking the columns of Y^* underneath each other. We can now write a linear model involving the random parameters, that is the elements of Ω_u, Ω_e, as follows

$$E(Y^{**}) = Z^*\theta \qquad (2.1.6)$$

where Z^* is the design matrix for the random parameters. An example of such a design matrix for a simple variance components model is given in Chapter 2. We now carry out a generalized least squares analysis to estimate θ, namely

$$\hat{\theta} = (Z^{*\mathrm{T}}V^{*-1}Z^*)^{-1}Z^{*\mathrm{T}}V^{*-1}Y^{**}, \quad V^* = V \otimes V \qquad (2.1.7)$$

where $\otimes$ is the Kronecker product. The covariance matrix of $\hat{\theta}$ is given by

$$(Z^{*\mathrm{T}}V^{*-1}Z^*)^{-1}Z^{*\mathrm{T}}V^{*-1}\,\mathrm{cov}(Y^{**})V^{*-1}Z^*(Z^{*\mathrm{T}}V^{*-1}Z^*)^{-1}$$

Now we have

$$Y^{**} = \mathrm{vec}(\tilde{Y}\tilde{Y}^{\mathrm{T}}) = \tilde{Y} \otimes \tilde{Y}$$

Using a standard result (for example Searle *et al*, 1992, section 12.3) we have

$$\mathrm{cov}(\tilde{Y} \otimes \tilde{Y}) = (V \otimes V)(I + S_N)$$

where $V \otimes V = V^*$ and S_N is the vec permutation matrix.

As Goldstein and Rasbash (1992) noted, the matrix A where $Z^* = \mathrm{vec}(A)$, is symmetric and hence

$$V^{*-1}Z^* = (V^{-1} \otimes V^{-1})\mathrm{vec}(A) = \mathrm{vec}(V^{-1}AV^{-1})$$

and $V^{-1}AV^{-1}$ is symmetric so that, using a standard result, we have

$$S_N V^{*-1}Z^* = V^{*-1}Z^*$$

and after substituting in the above expression for $\mathrm{cov}(\hat{\theta})$ we obtain

$$\text{cov}(\hat{\theta}) = 2(Z^{*T}V^{*-1}Z^*)^{-1} \tag{2.1.8}$$

The iterative generalized least squares (IGLS) procedure (Goldstein, 1986) iterates between (2.1.4) and (2.1.7) using the current estimates of the fixed and random parameters. Typical starting values for the fixed parameters are those from an ordinary least squares analysis. At convergence, assuming multivariate Normality, the estimates are maximum likelihood.

The IGLS procedure produces biased estimates in general and this can be important in small samples. Goldstein (1989a) shows how a simple modification leads to restricted iterative generalized least squares (RIGLS) or restricted maximum likelihood (REML) estimates which are unbiased. If we rewrite (2.1.5) using the *estimates* of the fixed parameters $\hat{\beta}$ we obtain

$$E(Y^*) = V_2 - X\,\text{cov}(\hat{\beta})X^{\text{T}} = V_2 - X(X^{\text{T}}V_2^{-1}X)^{-1}X^{\text{T}} \tag{2.1.9}$$

By taking account of the sampling variation of the $\hat{\beta}$ we can obtain an unbiased estimate of V_2 by adding the second term in (2.1.9), the 'hat' matrix, from Y^* at each iteration until convergence. In the case where we are estimating a variance from a simple random sample this becomes the standard procedure for using the divisor $n-1$, rather than n, to produce an unbiased estimate.

Full details of efficient computational procedures for carrying out all these calculations are given by Goldstein and Rasbash (1992).

APPENDIX 2.2

Multilevel residuals estimation

Denote the set of m_h residuals at level h in a multilevel model by

$$p_h = \{p_{h1}, \ldots, p_{hm_h}\}, \quad p_{hi}^T = \{p_{hi1}, \ldots, p_{hin_h}\} \tag{2.2.1}$$

where n_h is the number of level h units. Since the residuals at any level are independent of those at any other level, for each residual vector we require the posterior or predicted residual estimates given by

$$\hat{p}_{hi} = E(p_{hi} \mid \tilde{Y}, V)$$

where $\tilde{Y} = Y - X\beta$. We consider the regression of the set of all residuals p_h on $\tilde{Y}$ which gives the estimator

$$\hat{p}_h = R_h^T V^{-1} \tilde{Y} \tag{2.2.2}$$

where R_h is block diagonal, each block corresponding to a level h unit, and for the jth block given by

$$Z_{(j)}^h \Omega_h$$

where $Z_{(j)}^h$ is the matrix of explanatory variables for the random coefficients at level h. We obtain consistent estimators by substituting sample estimates of the parameters in (2.2.2). These estimates are linear functions of the responses and their unconditional covariance matrix is given by

$$R_h^T V^{-1} (V - X(X^T V^{-1} X)^{-1} X^T) V^{-1} R_h \tag{2.2.3}$$

The second term in (2.2.3) derives from considering the sampling variation of the estimates of the fixed coefficients and can be ignored in large samples, and we obtain a consistent estimator by substituting parameter estimates in

$$R_h^T V^{-1} R_h$$

Note that there are no covariances across units. Where we wish to study the distributional properties of standardized residuals for diagnostic purposes then the unconditional covariance matrix (2.2.3) should be used to standardize the estimated residuals. If, however, we wish to make inferences about the true p_{hi}, for example to construct confidence intervals or test differences, then we require the conditional covariance matrix of $\hat{p}_h \mid p_h$ or $E[(\hat{p}_h - p_h)(\hat{p}_h - p_h)']$ which is given by substituting parameter estimates in

$$S_h - R_h^T V^{-1}(V - X(X^T V^{-1} X)^{-1} X^T)V^{-1} R_h \qquad (2.2.4)$$

where S_h is the block diagonal matrix where each block corresponding to a level h unit is Ω_h. We note that no account is taken of the sampling variability associated with the estimates of the random parameters in (2.2.3) or (2.2.4). Thus, with small numbers of units, a procedure such as bootstrapping should be used to estimate these covariance matrices (Chapter 3).

APPENDIX 2.3

The EM algorithm

To illustrate the procedure, consider the 2-level variance components model

$$y_{ij} = (X\beta)_{ij} + u_j + e_{ij}, \quad \text{var}(e_{ij}) = \sigma_e^2, \quad \text{var}(u_j) = \sigma_u^2 \tag{2.3.1}$$

The vector of level 2 residuals is treated as missing data, and the 'complete' data therefore consists of the observed vector Y and the u_j treated as observations. The joint distribution of these, assuming Normality, and using our standard notation is

$$\begin{bmatrix} Y \\ u \end{bmatrix} = N \left\{ \begin{bmatrix} X\beta \\ 0 \end{bmatrix}, \begin{bmatrix} V & J^T\sigma_u^2 \\ \sigma_u^2 J & \sigma_u^2 I \end{bmatrix} \right\} \tag{2.3.2}$$

This readily generalizes to the case where there are several random coefficients. If we denote these by β_j we note that some of them may have zero variances. We can now derive the distribution of $\beta_j | Y$ in Appendix 2.2, and we can also write down the log-likelihood function for (2.3.2) with a general set of random coefficients, namely

$$\log(L) \propto -N \log(\sigma_e^2) - J \log|\Omega| - \sigma_e^{-2} \sum_{ij} e_{ij}^2 - \sum_j \beta_j^T \Omega_u^{-1} \beta_j \tag{2.3.3}$$

$$\Omega_u = \text{cov}(\beta_j)$$

Maximizing this for the random parameters we obtain

$$\hat{\sigma}_e^2 = N^{-1} \sum_{ij} e_{ij}^2$$

$$\hat{\Omega}_u = J^{-1} \sum_j \beta_j \beta_j^T \tag{2.3.4}$$

where J is the number of level 2 units. We do not know the values of the individual random variables. We require the expected values, conditional on the Y and the current parameters, of the terms under the summation signs—these being the sufficient statistics. We then substitute these expected values in (2.3.4) for the updated random parameters. These conditional values are based upon the 'shrunken' predicted values and their (conditional) covariance matrix, given in Appendix 2.2. With these updated values of the random parameters we can form V and hence obtain the updated estimates for the fixed parameters using generalized least squares. We note that the expected values of the sufficient statistics can be obtained using the general result for a random parameter vector θ

$$E(\theta\theta^{T}) = \text{cov}(\theta) + [E(\theta)][E(\theta)]^{T} \tag{2.3.5}$$

The prediction is known as the E step of the algorithm, and the computations in (2.3.4) the M step. Given starting values, based upon OLS, these computations are iterated until convergence is obtained. Convenient computational formulae for computing these quantities at each iteration can be found in Bryk and Raudenbush (1992).

Using the general procedures for estimating residuals in Appendix 2.2, at each iteration we would define the estimated residuals as explanatory variables and then regress the response variable on these. In the present case this would be an OLS regression to obtain the fixed coefficients. Note, however, that we require the matrix given by (2.3.5) in the estimation rather than the usual $(\theta^{T}V^{-1}\theta) \propto (\theta^{T}\theta)$ which in this case is just the second term in (2.3.5), the first term being the (estimated) covariance matrix of the residuals. Using (2.3.4) for the random parameters we then estimate new residuals and iterate.

APPENDIX 2.4

Gibbs sampling

Gibbs sampling is a procedure belonging to the set of Markov Chain Monte Carlo algorithms, which exploits the properties of Markov chains where the probability of an event is conditionally dependent on a previous state. The procedure is iterative and at each stage, from the full multivariate distribution, the distribution of each component conditional on the remaining components is computed and used to generate a random variable. The components may be variates, regression coefficients, covariance matrices etc. After a suitable number of iterations, we obtain a sample of values from the distribution of any component which we can then use to derive any desired characteristic such as the mean, covariance matrix, etc. Gilks *et al* (1993) provided a comprehensive discussion with applications, and an application to a 2-level logit model is given by Zeger and Karim (1991).

We outline the procedure for a 2-level model.

Write

$$Y = X\beta + Z^{(2)}u + Z^{(1)}e$$

We first consider the distribution $\beta \mid u^{(k)}, Y$ where k refers to the kth iteration.

Given $u^{(k)}$, $Z^{(2)}u$ is just an offset so that we can regress y_{ij} on x_{ij} to estimate $\hat{\beta}^{(k)}$ and $\text{var}(\hat{\beta}^{(k)})$.

We can then select a random vector from this distribution, assumed to be multivariate normal $(\hat{\beta}^{(k)}, \text{var}(\hat{\beta}^{(k)}))$.

We now consider the distribution of $\Omega_2 \mid u^{(k)}$. We have (with a non-informative prior) that the (posterior) distribution of Ω_2^{-1} is a Wishart distribution with parameter (i.e. covariance) matrix

$$S^{(k)} = \sum_{j=1}^{J} u_j^{(k)} u_j^{(k)^{\mathrm{T}}} \text{ with } d = J - q + 1 \text{ d.f.}$$

where J is the number of level 2 units and q is the number of random coefficients.

A simple way of generating such a Wishart distribution is to generate d multivariate normal vectors from $N(0, S^{(k)})$ and form their SSP matrix. This provides $\Omega_2^{(k)}$.

Finally, we consider the distribution $u_j \mid \beta, \Omega_2, Y$. These are the usual level residuals, for which we have standard expressions for their expected values and covariance matrix. We note that for a 2-level model (but not within a 3-level model these are block-independent. Assuming Normality, we can now generate a set of $u_j^{(k)}$ and this completes an iterative cycle.

There are some particular computational details to be noted. For example 'rejection sampling' at each cycle can be used and we can do several cycles for Ω_2, u_j for each β since the former tend to have higher autocorrelations across cycles.

The procedure can be applied to any existing models, e.g. logit models, where the conditional distributional assumptions are explicit. Gibbs Sampling tends to be computationally demanding, with hundreds if not thousands of iterations required explored for their fit to the data. It is perhaps most useful for small and moderate sized samples and when used in conjunction with likelihood-based EM or IGLS algorithms.

3

Extensions to the Basic Multilevel Model

Complex variance structures

3.1   In all the models of Chapter 2 we have assumed that a single variance describes the random variation at level 1. At level 2 we have introduced a more complex variance structure, as shown in Fig. 2.7, by allowing regression coefficients to vary across level 2 units. The modelling and interpretation of this complex variation, however, was solely in terms of randomly varying coefficients. Now, we look at how we can model the variation explicitly as a function of explanatory variables and how this can give substantively interesting interpretations. We shall consider mainly the level 1 variation, but the same principles apply to higher levels. We shall also, in this chapter, consider extensions of the basic model to include constraints on parameters, unit weighting, standard error estimation and aggregate level analyses.

In the analysis of the JSP data in Chapter 2 we saw that the level 1 residual variation appeared to decrease with increasing 8-year maths score. We also saw how the estimated individual school lines appeared to converge at high 8-year scores. We consider first the general problem of modelling the level 1 variation.

Since we shall now consider several random variables at each level, the notation used in Chapter 2 needs to be extended. For a 2-level model we continue to use the notation u_j, e_{ij} for the total variation at levels 2 and 1 and we write

$$u_j = \sum_{h=0}^{r_2} u_{hj} z_{hj}, \quad e_{ij} = \sum_{h=0}^{r_1} e_{hij} z_{hij} \tag{3.1}$$

where the zs are explanatory variables. Normally z_{0j}, z_{0ij} refer to the constant (=1) defining a basic or intercept variance term at each level.

For 3-level models we will use the notation v_k, u_{kj}, e_{ijk} where i refers to level 1 units, j to level 2 units, and k to level 3 units and h indexes the explanatory variables and their coefficients within each level.

One simple model for the level 1 variation is to make it a linear function of a simple explanatory variable. Consider the following extension of (2.1)

$$y_{ij} = \beta_0 + \beta_1 x_{ij} + (u_j + e_{0ij} + e_{1ij}z_{ij}), \quad z_{ij} = x_{ij}$$
$$\left. \text{var}(e_{0ij}) = \sigma_{e0}^2, \quad \text{var}(e_{1ij}) = 0, \quad \text{cov}(e_{0ij}e_{1ij}) = \sigma_{e01} \right\} \quad (3.2)$$

so that the level 1 contribution to the overall variance is the linear function of z_{ij}

$$\sigma_{e0}^2 + 2\sigma_{e01}z_{ij}$$

This device of constraining a variance parameter to be zero in the presence of a non-zero covariance is used to obtain the required variance structure. Thus, it is only the specified *functions* of the random parameters in (3.2) which have an interpretation in terms of the level 1 variances of the responses y_{ij}. This will generally be the case where the coefficients are random at the same level at which the explanatory variables are defined. Thus, for example, in the analyses of the JSP data in Chapter 2, we could model the average school 8-year score, which is a level-2 variable, as random at level 2. If the resulting variance and covariance are non-zero, the interpretation will be that the between-school variance is a quadratic function of the 8-year score, namely

$$\sigma_{u0}^2 + 2\sigma_{u01}z_j + \sigma_{u1}^2 z_j^2$$

where z_j is the average 8-year score.

Furthermore, we can allow a variance parameter to be negative, so long as the total level 1 variance remains positive within the range of the data. In Chapter 5 we discuss modelling the total level 1 variance as a nonlinear function of explanatory variables; for example, as a negative exponential function which automatically constrains the variance to be positive.

Where a coefficient is made random at a level higher than that at which the explanatory variable itself is defined, then the resulting variance (and covariance) can be interpreted as the between-higher-level unit variance of the within-unit relationship described by the coefficient. This is the interpretation, for example, of the random coefficient model of Table 2.5 where the coefficient of the student 8-year score varies randomly across schools. In addition, of course, we have a complex variance (and covariance) structure at the higher level.

The model (3.2) does not constrain the overall level 1 contribution to the variance in any way. In particular, it is quite possible for the level 1 variance and hence the *total* response variance to become negative. This is clearly inadmissible and will also lead to numerical estimation problems. To overcome this we can consider elaborating the model by adding a quadratic term, most simply by removing the zero constraint on the variance. In Chapter 5, we consider the alternative of modelling the variance as a nonlinear function of explanatory variables.

In Table 3.1 we extend the model of Table 2.5 to incorporate such a quadratic function for the level 1 variance. If we attempt to fit a linear function we indeed find that a negative total variance is predicted.

The results from model A show a significant complex level 1 variation (chi-squared with 2 degrees of freedom = 123). Furthermore, the level 2 correlation between the intercept and slope is now reduced to -0.91 and with little change among the fixed part coefficients. The predicted level 1 standard deviation varies from about 9.0 at the lowest 8-year score value to about 1.9 at the highest, reflecting the impression from the scatterplot in Fig. 2.1.

One of the reasons for the high negative correlation between the intercept and

Table 3.1 JSP data with level 1 variance a quadratic function of 8-year score measured about the sample mean. Model A with original scale; models B and C with Normal score transform of 11-year score

Parameter	Estimate (s.e.) A	Estimate (s.e.) B	Estimate (s.e.) C
Fixed:			
Constant	31.7	0.13	0.14
8-year score	0.58 (0.029)	0.097 (0.004)	0.096 (0.004)
Gender (boys–girls)	−0.35 (0.26)	−0.04 (0.05)	−0.03 (0.05)
Social class (Non Man.–Man.)	0.74 (0.29)	0.16 (0.06)	0.16 (0.06)
School mean 8-year score	0.02 (0.11)	−0.008 (0.02)	
8-yr score × school mean 8-yr score	0.02 (0.01)	0.0006 (0.02)	
Random:			
Level 2			
σ^2_{u0}	2.84 (0.88)	0.084 (0.024)	0.086 (0.024)
σ_{u01}	−0.17 (0.07)	−0.0024 (0.0015)	−0.0030 (0.0015)
σ^2_{u1}	0.012 (0.007)	0.00018 (0.00016)	0.00021 (0.00016)
Level 1			
σ^2_{e0}	16.5 (1.02)	0.413 (0.029)	0.412 (0.022)
σ_{e01}	−0.90 (0.02)	−0.0032 (0.0017)	
σ^2_{e1}	0.06 (0.02)	0.0000093 (0.00041)	

slope at the school level may be associated with the fact that the 11-year score has a 'ceiling' with a third of the students having scores of 35 or more out of 40. A standard procedure for dealing with such skewed distributions is to transform the data, for example to normality, and this is most conveniently done by computing Normal scores; that is by assigning Normal order statistics to the ranked scores. The results from this analysis are given under model B in Table 3.1. Note that the scale has changed since the response is now a standard normal variable with zero mean and unit standard deviation. We now find that there is no longer any appreciable complex variation at level 1; the chi-squared test yields a value of 3.4 on 2 degrees of freedom. Nor is there any effect of the compositional variable of the mean school 8-year score; the chi-squared test for the two fixed coefficients associated with this give a value of 0.2 on 2 degrees of freedom. The reduced model is fitted as C. The parameters associated with the random slope at level 2 remain significant ($\chi^2_2 = 7.7, P = 0.02$) and the level 2 correlation is further reduced to −0.71. Figure 3.1 shows the level 1 standardized residuals plotted against the predicted values, from which it is clear that now the variance is much more nearly constant. This example demonstrates that interpretations may be sensitive to the scale on which variables are measured. It is typical of many measurements in the social sciences that their scales are arbitrary and we can justify nonlinear, but monotone order preserving, transformations if they help to simplify the statistical model and the interpretation.

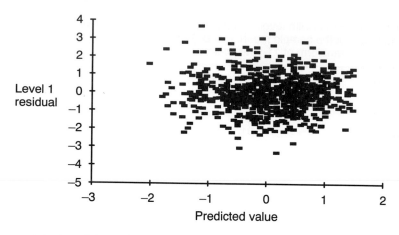

Figure 3.1   Level 1 standardized residuals by predicted values for analysis C in Table 3.1.

We are not limited to making the variance a function of a single explanatory variable, and we can consider general functions of these combined. Some may be absent from the fixed part of the model, or equivalently have their fixed coefficients constrained to zero. A traditional, single level, example is 'regression through the origin' in which the fixed intercept term is zero while a level 1 variance associated with the intercept is fitted.

We can consider any particular function of explanatory variables as the basis for modelling the variance. One possibility is to take the fixed part predicted value $\hat{y}_{ij}$ and define the level 1 random term as $e_{1ij}\sqrt{\hat{y}_{ij}}$, assuming the predicted value is positive, so that the level 1 variance becomes $\sigma_{ei}^2\hat{y}_{ij}$; that is, proportional to the predicted value—often known as a 'constant coefficient of variation' model. Other functions are clearly possible and, as we shall see in Chapter 7, often there are natural choices associated with distributional assumptions made about the responses.

Variances for subgroups defined at level 1

3.1.1   A common example of complex variation at level 1 is where variances are specific for subgroups. For example, for many measurements there are gender or social class differences in the level 1 variation. A straightforward way to model this situation in the case of a single such grouping is by defining the following version of (3.2) for a model with different variances for children with manual and with non-manual social class backgrounds.

$$
\left.
\begin{aligned}
y_{ij} &= \beta_0 + \beta_1 x_{ij} + (u_{0j} + e_{2ij}z_{2ij} + e_{3ij}z_{3ij}) \\
z_{2ij} &= 1 \text{ for manual, } 0 \text{ for non-manual} \\
z_{3ij} &= 0 \text{ for manual, } 1 \text{ for non-manual} \\
\text{var}(e_{2ij}) &= \sigma_{e2}^2, \quad \text{var}(e_{3ij}) = \sigma_{e3}^2, \quad \text{cov}(e_{2ij}, e_{3ij}) = 0
\end{aligned}
\right\}
\quad (3.3)
$$

Table 3.2 JSP data with normal score of 11-year maths as response. Subscript 1 refers to 8-year maths score, 2 to manual group, 3 to non-manual group and 4 to boys

Parameter	Estimate (s.e.) A	Estimate (s.e.) B	Estimate (s.e.) C
Fixed:			
Constant	0.13	0.13	0.13
8-year score	0.096 (0.004)	0.096 (0.004)	0.096 (0.004)
Gender (boys–girls)	−0.03 (0.05)	−0.03 (0.05)	−0.03 (0.05)
Social Class (Non Man.–Man.)	0.16 (0.05)	0.16 (0.05)	0.16 (0.05)
Random:			
Level 2			
σ_{u0}^2	0.086 (0.025)	0.086 (0.025)	0.086 (0.024)
σ_{u01}	−0.0029 (0.0015)	−0.0029 (0.0015)	−0.0028 (0.0015)
σ_{u1}^2	0.00018 (0.00015)	0.00018 (0.00015)	0.00018 (0.00015)
Level 1			
σ_{e0}^2		0.37 (0.04)	0.36 (0.04)
σ_{e02}		0.03 (0.02)	0.03 (0.02)
σ_{e2}^2	0.43 (0.03)		
σ_{e3}^2	0.37 (0.04)		
σ_{e04}			0.004 (0.02)
−2 (log-likelihood)	1491.8	1491.8	1491.7

If we do this for model C in Table 3.1 then we obtain the estimates in column A of Table 3.2.

The estimates of the fixed parameters have changed little and the level 2 parameters are also similar. At level 1 the variance for the manual students is higher than that for the non-manual students, but not significantly since the likelihood ratio test statistic, formed by differencing the values of (−2 log-likelihood) for the model with a single level 1 variance (1493.7) and that given in analysis A of Table 3.2, gives a chi-squared test statistic of 1.9 on 1 degree of freedom.

We now look at an alternative method for specifying this type of complex variation at level 1, which has certain advantages. We write

$$y_{ij} = \beta_0 + \beta_1 x_{ij} + (u_{0j} + e_{0ij} + e_{2ij} z_{2ij})$$
$$z_{2ij} = 1 \text{ for manual, } 0 \text{ for non-manual}$$
$$\text{var}(e_{0ij}) = \sigma_{e0}^2, \quad \text{var}(e_{2ij}) = 0, \quad \text{cov}(e_{0ij}, e_{2ij}) = \sigma_{e02}$$

and the level 1 variance is given by $\sigma_{e0}^2 + 2\sigma_{e02}z_{2ij}$ because we have constrained the variance of the manual coefficient to be zero. Thus, for manual children ($z_{2ij} = 1$) the level 1 variance is $\sigma_{e0}^2 + 2\sigma_{e02}$ and for non-manual children the level 1 variance is σ_{e0}^2. The second column in Table 3.2 gives the results from this formulation and we see that, as expected, the covariance estimate is equal to half the difference between the separate variance estimates in the first column.

Suppose now that we wish to model the level 1 variance as a function of both social class group and gender. One possibility is to fit a separate variance for each of the 4 possible resulting groups, using either of the above procedures. Another possibility is to consider a more parsimonious 'additive' model for the variances as follows

$$
\left.
\begin{aligned}
e_{ij} &= e_{0ij} + e_{2ij}z_{2ij} + e_{4ij}z_{4ij} \\
z_{4ij} &= 1 \text{ if a boy, } 0 \text{ if a girl} \\
\mathrm{var}(e_{0ij}) &= \sigma_{e0}^2, \quad \mathrm{cov}(e_{0ij}e_{2ij}) = \sigma_{e02}, \quad \mathrm{cov}(e_{0ij}, e_{4ij}) = \sigma_{e04}
\end{aligned}
\right\} \qquad (3.4)
$$

with the remaining two variances and covariance equal to zero. Thus, (3.4) implies that the level 1 variance for a manual boy is $\sigma_{e0}^2 + 2\sigma_{e02} + 2\sigma_{e04}$ etc. The third column of Table 3.2 gives the estimates for this model and we see that there is a negligible difference in the level 1 variance for boys and girls.

We can extend such structuring to the case of multicategory variables and we can also include continuous variables as in Table 3.1. Suppose we had a 3 category variable: we define two dummy variables, say z_{5ij}, z_{6ij} corresponding to the second and third categories, just as if we were fitting the factor in the fixed part of the model. With z_{1ij} representing the continuous variable an additive model for the level 1 random variation can be written as

$$
e_{ij} = e_{0ij} + e_{5ij}z_{5ij} + e_{6ij}z_{6ij}
$$
$$
\mathrm{var}(e_{0ij}) = \sigma_{e0}^2, \quad \mathrm{var}(e_{1ij}) = \sigma_{e1}^2, \quad \mathrm{cov}(e_{0ij}e_{1ij}) = \sigma_{e01}
$$
$$
\mathrm{cov}(e_{0ij}e_{5ij}) = \sigma_{e05}, \quad \mathrm{cov}(e_{0ij}e_{6ij}) = \sigma_{e06}
$$

This model can be elaborated by including one or both of the covariances between the dummy variable coefficients and the continuous variable coefficient, namely σ_{e15}, σ_{e16}. These covariances are analogous to interaction terms in the fixed part of the model and we see that, starting with an additive model, we can build up models of increasing complexity. The only restriction is that we cannot fit covariances between the dummy variable categories for a single explanatory variable. Thus, if social class had three categories, we could fit two covariances corresponding to, say, categories 2 and 3 but not a covariance *between* these categories.

Residuals can be estimated in a straightforward manner for these complex variation models. For example, from (3.4) the estimated residual for a manual boy is $\hat{e}_{0ij} + \hat{e}_{2ij} + \hat{e}_{4ij}$ where the estimates of the individual residuals are computed using the formulae in Appendix 2.2 with the appropriate zero variances.

Variance as a function of predicted value

3.1.2   The level 1 variance can be modelled as a function of any combination of explanatory variables and, in particular, we can incorporate the estimated coefficients themselves in such functions. A useful special case is where the function is the fixed part predicted value $\hat{y}_{ij}$. Thus (3.2) becomes

$$
y_{ij} = \beta_0 + \beta_1 x_{ij} + (u_{0j} + e_{0ij} + e_{1ij}\hat{y}_{ij})
$$

Table 3.3 GCSE scores related to secondary school intake achievement

	A	B
Fixed:		
Constant	0.13	0.14
Reading score	0.50 (0.03)	0.49 (0.03)
Gender (boys–girls)	−0.19 (0.06)	−0.22 (0.06)
Social class (Non Man.–Man.)	−0.07 (0.06)	−0.06 (0.06)
Random:		
Level 2		
σ_{u0}^2	0.03 (0.02)	0.02 (0.01)
Level 1		
σ_{e0}^2	0.66 (0.04)	0.63 (0.04)
σ_{e01}		0.16 (0.04)
σ_{e1}^2		0.11 (0.09)
−2 (log-likelihood)	1929.5	1905.0

with level 1 variance given by $\sigma_{e0}^2 + 2\sigma_{e01}\hat{y}_{ij} + \sigma_{e1}^2\hat{y}_{ij}^2$. A special case of this model is the so-called 'constant coefficient of variation model' where the two variance terms are constrained to zero. The estimation of the random parameters is straightforward: at each iteration of the algorithm a new set of predicted values is calculated and used as the level 1 explanatory variable.

Table 3.3 illustrates the use of this model where the level 1 variance shows a strong dependence on the predicted value. The data are the General Certificate of Secondary Education (GCSE) scores at the age of 16 years of the Junior School Project students. This score is derived by assigning values to the grades achieved in each subject examination and summing these to produce a total score (see Nuttall *et al*, 1989 for a detailed description). There are 785 students in this analysis in 116 secondary schools to which they transferred at the age of 11 years. The students have a measure of reading achievement, the London Reading Test (LRT) taken at the end of their junior school, and this is used as a pretest baseline measure against which relative progress is judged. Both the reading test score and the examination score have been transformed to Normal equivalent deviates.

Analysis A is a variance components analysis and Fig. 3.2 shows a plot of the standardized level 1 residuals against the predicted values. It is clear that the variation is much smaller for low predicted values.

One possible extension of the model to deal with this is to use the LRT score as an explanatory variable at level 1, so that the level 1 variance becomes a quadratic function of the LRT score. This does not, however, entirely eliminate the relationship and instead we model the predicted value as a level 1 explanatory variable, and the results are presented as analysis B of Table 3.3. If we now plot the standardized residuals associated with the intercept against the predicted values we obtain the pattern in Fig. 3.3 from which it is clear that much of the relationship between the variance and the predicted value has been accounted for. We could go on to fit more complex functions of the predicted value, for example involving nonlinear or higher order polynomial terms.

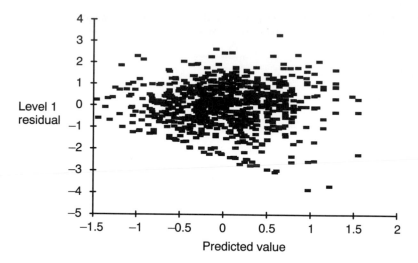

Figure 3.2   Standardized residuals for variance components analysis.

Variances for subgroups defined at higher levels

3.1.3   The random slopes model in Table 3.1 has already introduced complex variation at level 2 when the coefficient of a level 1 explanatory variable is allowed to vary across level 2 units. Just as with level 1 complex variation, we can also allow coefficients of variables defined at level 2 to vary at level 2. Exactly the same considerations apply for categorical level 2 variables as we had for such variables at level 1 and complex additive or interactive structures can be defined.

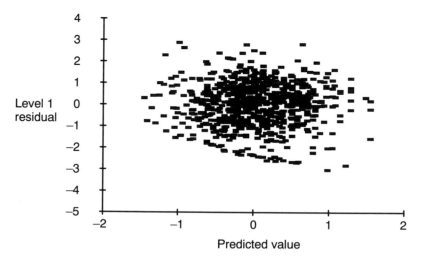

Figure 3.3   Standardized residuals with level 1 variance a function of predicted value.

In addition, the coefficient of a level 2 variable can vary randomly at either level 1 or level 2 or both. For example, suppose we have three types of school: all boys schools, all girls schools and mixed schools. We can allow different variances, at level 2, between boys schools, between girls schools and between mixed schools. We can also allow different between-student variances for each type of school.

To further illustrate complex level 1 variation and also to introduce a three-level model we turn to another data set, this time from a survey of social attitudes.

A 3-level complex variation model

3.2 The longitudinal or panel data come from the British Social Attitudes Survey and cover the years 1983–1986 with a random sample of 264 adults measured a year apart on four occasions and living at the same addresses. This panel was a sub-sample of a larger series of cross-sectional surveys. The final sample was intended to be self-weighting with each household, as represented by a single person, having the same inclusion probability. A full technical account of the sampling procedures is given by McGrath and Waterton (1986). The sampling procedure was, at the first stage, to sample parliamentary constituencies with probability proportional to size of electorate, then to sample a single 'polling district' within each constituency in a similar way and finally to sample an equal number of addresses within each polling district.

Because only one polling district was sampled from each constituency, we cannot separate the between-district from the between-constituency variation; the two are 'confounded'. Likewise, we cannot separate the between-individuals from the between-households variation. The basic variation is therefore at two levels, between-districts (constituencies) and between-individuals (households). The longitudinal structure of the data, with four occasions, introduces a further level below these two, namely a between-occasion-within-individual level, so that occasion is level 1, individual is level 2 and district is level 3. In Chapter 5 we shall study longitudinal data structures in more depth, at both level 1 and higher levels.

The response variable we shall use is a scale, in the range 0–7, concerned with attitudes to abortion. It is derived by summing the (0,1) responses to seven questions and can be interpreted as indicating whether the respondent supported or opposed a woman's right to abortion with high scores indicating strong support. Explanatory variables are political party allegiance (four categories), self-assessed social class (three categories), gender, age (continuous), and religion (four categories) and year (four categories). A number of preliminary analyses have been carried out and the effects of party allegiance, social class, gender and age, were found to be small and not statistically significant. We therefore examine the basic 3-level model, which can be written as follows

$$y_{ijk} = \beta_0 + (\beta_1 x_{1ijk} + \beta_2 x_{2ijk} + \beta_3 x_{3ijk})$$
$$+ (\beta_4 x_{4ijk} + \beta_5 x_{5ijk} + \beta_6 x_{6ijk}) + (v_k + u_{jk} + e_{ijk}) \qquad (3.5)$$

with the explanatory variables with subscripts 1–3 being dummy variables for religious categories 2–4 and those with subscripts 4–6 being dummy variables for years 1984–1986. We have three variances, one at each level in the random part of the model. The response variable in the following analyses has only eight cate-

Table 3.4 Repeated measurements of attitudes to abortion. Response is Normal score transformation. Religion estimates are contrasted with none. Age is measured about the mean of 37 years

Parameter	Estimate (s.e.) A	Estimate (s.e.) B	Estimate (s.e.) C
Fixed:			
Constant	0.32	0.33	0.33
Religion: R. Catholic	−0.80 (0.18)	−0.80 (0.18)	−0.69 (0.18)
Protestant	−0.27 (0.10)	−0.26 (0.10)	−0.25 (0.10)
Other	−0.63 (0.13)	−0.63 (0.13)	−0.54 (0.14)
Year: 1984	−0.29 (0.05)	−0.29 (0.48)	−0.29 (0.05)
1985	−0.06 (0.05)	−0.07 (0.05)	−0.07 (0.05)
1986	0.06 (0.05)	0.05 (0.04)	0.05 (0.04)
Age			0.013 (0.005)
Age × R. Catholic			−0.036 (0.010)
Age × Protestant			−0.014 (0.007)
Age × Other			−0.023 (0.008)
Random:			
Level 3			
σ_v^2	0.03 (0.02)	0.03 (0.02)	0.03 (0.02)
Level 2			
σ_u^2	0.37 (0.04)		0.34 (0.04)
Level 1			
σ_{e0}^2	0.31 (0.02)	0.21 (0.08)	0.21 (0.03)
σ_{e01}		0.11 (0.05)	0.10 (0.04)
σ_{e02}		0.03 (0.16)	0.03 (0.02)
σ_{e03}		0.04 (0.02)	0.04 (0.02)
σ_{e04}		0.05 (0.02)	0.05 (0.02)
σ_{e05}		0.05 (0.02)	0.05 (0.02)
σ_{e06}		0.00 (0.02)	0.00 (0.02)
−2 (log-likelihood)	2233.5	2214.2	2198.7

gories, with 32% of the sample having the highest value of 7. The response has been transformed by assigning Normal scores to the overall distribution and we shall treat the response as if it was continuously distributed. In Chapter 7 we shall look at other models which retain the categorization of the response variable.

Table 3.4 gives the results of fitting (3.5). The between-occasion and between-individual variances are similar. The level 3 variance is small, and the likelihood ratio chi-squared is 2.05 (compared with a value of 1.64 obtained from comparing the estimate with its standard error), which is not significant at the 10% level.

For the religious differences we have $\chi_3^2 = 33.7$ for the overall test with all those having religious beliefs being less inclined to support abortion, the Roman Catholic and other religions being least likely of all. The Roman Catholic and other religions are significantly less likely than the Protestants to support abortion. The simultaneous test (3 d.f.) chi-squared statistics respectively are 9.7 and 9.0 ($P = 0.03$). For the year differences we have $\chi_3^2 = 59.7$ and simultaneous comparisons show that in 1984 there was a substantially less approving attitude towards abortion. It is likely that this is an artefact of the way the questions were put to respondents.[1] There is no significant interaction between religion and year.

We now look at elaborating the random structure of the model. At level 1 we fit an additive model, as in the section on variances for subgroups defined at level 1, for the categories of religion and for year. Year is the variable defining level 1, but religion is defined at level 2 and is an example of a higher level variable used to define complex variation at a lower level.

The results are given as analysis B in Table 3.3. For year we obtain $\chi_3^2 = 8.3$ ($P = 0.04$) and for religion $\chi_3^2 = 11.0$ ($P = 0.01$). There is a greater heterogeneity within the Roman Catholics, from year to year, and within the other religions than within Protestants and those with no religion. The addition of these variances to the model does not change substantially the values for the other parameters.

Fitting complex variation at level 2 (between individuals) and level 3 (between districts) does not yield statistically significant effects, although there is some suggestion that there may be more variation among Roman Catholics.

For the final analysis we look again at the fixed part and explore interactions. None of the interactions have important effects except for that of age with religion, although age on its own had a negligible effect. We see from analysis C that those with no religion show an increasing approval of abortion with age, whereas the Roman Catholics, and to a smaller extent other religions, show a decreasing approval with age. The overall chi-squared for testing the interactions is 16.1 with 3 degrees of freedom.

Parameter constraints

3.3   In the example of the previous section some of the fixed and random parameters for year and religious groups were similar. This suggests that we could fit a simpler model by forcing or 'constraining' such parameters to take the same values and so decreasing the standard errors in the model. We illustrate the procedure using the fixed part estimates for the abortion attitudes data.

We consider the general linear constraint for the fixed parameters in the form $C\beta = k$, where C is an $(n \times p)$ constraint matrix and k is a vector which can have quite general values for its elements.

Suppose that, in analysis C of Table 3.4, we wished to constrain the main effects and interaction terms of the Roman Catholic and Other religions to be equal. This implies two constraint functions, and we have

$$C = \begin{pmatrix} 0 & 1 & 0 & -1 & 0 & 0 & 0 & 0 & 0 & 0 & 0 \\ 0 & 0 & 0 & 0 & 0 & 0 & 0 & 0 & 1 & 0 & -1 \end{pmatrix}$$

$$k = \begin{pmatrix} 0 \\ 0 \end{pmatrix}$$

which implies $\hat{\beta}_1 = \hat{\beta}_3$, $\hat{\beta}_8 = \hat{\beta}_{10}$.

The constrained estimator of β is

$$\hat{\beta}^c = \hat{\beta} - LC(C^{\mathrm{T}}LC)^{-1}(C^{\mathrm{T}}\hat{\beta} - k)$$

$$L = (X^{\mathrm{T}}\hat{V}^{-1}X)^{-1}$$

(3.6

where $\hat{\beta}$ is the unconstrained estimator. The covariance matrix of the constrained estimator is *MLM* where

$$M = I - LC(C^{\mathrm{T}}LC)^{-1}C^{\mathrm{T}}$$

There is an analogous formula for constrained random parameters.

Using the above constraints for analysis C in Table 3.4, the random parameters are little changed, the main effects for Roman Catholic and Other religions become −0.57, the interaction terms become −0.026 and the remaining main effects are virtually unaltered. The standard errors, as expected, are smaller, being 0.121 for the main effect estimate and 0.007 for the interaction.

In addition to linear constraints we can also apply nonlinear constraints. To illustrate the procedure we consider the analysis in Table 2.5, where the estimated correlation between the slope and intercept was −1.03. To constrain this to be exactly −1.0, after each iteration of the algorithm we compute the covariance as a function of the variances to give this correlation. Thus, after iteration t we compute $\sigma_{u01}^{t+1} = \hat{\sigma}_{u0}^t \hat{\sigma}_{u1}^t$ and then constrain the covariance to be equal to this value, a linear constraint, for iteration $t+1$. This procedure is repeated until convergence is obtained for the unconstrained values. For more general nonlinear constraints we may require several such constraints to apply simultaneously.

If we constrain the model of Table 2.5 to give a correlation of −1.0 we find that the fixed effects and the level 1 variance are altered only slightly, with a small reduction in standard errors. The level 2 parameters, however, are reduced by about 50% and are closer to those in analysis A of Table 3.1 where the estimated correlation is −0.91.

We can also temporarily constrain values during the iterative estimation procedure if convergence is difficult or slow. Some parameters, or functions of them, can be held at current values, other parameter values allowed to converge and the constrained parameters subsequently unconstrained.

Weighting units

3.4   It is common in sample surveys to select level 1 units, for example household members, so that each unit in the population has the same probability of selection. Such self-weighting samples can then be modelled using any of the multilevel models of this book. Likewise, if the model correctly specifies the population structure, non-self-weighting samples can be modelled similarly: the differential selection probabilities contain no extra information for the model parameters. If we wished to form predictions for the whole population on the basis of the model estimates, we could apply the weight (typically the inverse of the selection probability) to the predicted value for each level 1 unit and then form a weighted sum over these units.

In some cases, however, the sample inclusion probabilities are designed to be equal, but there is differential non-response. It is common in such cases to assign weights to the level 1 units to compensate for the non-response, although we shall not here deal with the case where non-response is associated with the values of the residuals. Another situation where weighting should be used is when the modelling of the population structure is inadequate, that is the model is misspecified. If we then wish to make inferences about population values which are robust against poor model specification we will need to weight the units. As a general procedure, therefore, to guard against model misspecification weighting is to be recommended. In Chapter 7 we give an example where the weighting of units in an incompletely specified model changes some of the parameter estimates.

In some data analyses, we may come across values which are possible errors. Rather than excluding the units containing these, we may wish to keep them but assign them a lower weighting in the analysis, in line with the extent to which we believe they are in error.

In a 2-level model (with extensions to higher levels being straightforward) define the level 1 weight vector

$$W = \{w_{ij}\}, \quad \sum_{ij} w_{ij} = 1$$

If we wish to assign differential weights w_j to level 2 units then we can assign the weights w_j/n_j, $\sum_j w_j = 1$, to each level 1 unit in the jth level 2 unit. We can also combine level 1 weights with these level 2 weights by defining a further set of weights for the level 1 units within each level 2 unit w'_{ij}, $\sum_i w'_{ij} = n_j$, with $w_{ij} = w'_{ij}w_j$.

The estimator of the fixed parameters becomes

$$\left.\begin{aligned} \hat{\beta}_W &= (X^T V_W^{-1} X)^{-1} X^T V_W^{-1} Y \\ V_W &= WVW \end{aligned}\right\} \tag{3.7}$$

where W is the diagonal matrix with elements $\sqrt{w_{ij}}$ on the diagonal. There is an analogous formula for the random parameters.

The covariance matrix of these estimates is given by

$$\mathrm{cov}(\hat{\beta}_W) = (X^T V_W^{-1} X)^{-1} X^T V_W^{-1} V V_W^{-1} (X^T V_W^{-1} X)^{-1}$$

When estimating residuals we may sometimes wish to use a different set of weights. In a 2-level model, weights may be applied to the level 1 units to reflect an overall sampling procedure, but we may wish to apply different weights to reflect the composition of each level 2 unit separately. For example, suppose we have schools with an equally weighted sample where every level 1 unit has the same probability of selection. The actual sample proportions of, say, male and female students may be very different from that in the school as a whole. If there are important gender differences then in such a case we may wish to apply weights which reflect the true rather than the sample proportions.

To estimate weighted residuals, in Equation (2.2.2) of Appendix 2 we replace V by V_W and R_h by $R_{h(W)}$ where

$$R_{h(W)} = R_h \overline{W}, \quad \overline{W} = \oplus_j \overline{W}_j I_{n_j}, \quad \overline{W}_j = \sum_j w_{ij}/n_j$$

with corresponding changes to the estimated covariance matrix.

Robust, jackknife and bootstrap uncertainty estimates

3.5   Until now we have assumed that the response variable has a Normal distribution, and where the departure from Normality is substantial we have considered a transformation, using Normal scores. As we saw in the abortion data set, however, such transformations may be only approximate where the original score distribution is highly discrete or very skew. The estimates of the fixed and random parameters will still be consistent when the Normality assumption is untrue, but the standard error estimates cannot be used to obtain confidence intervals or to test significance except in large samples.

One way of attempting to deal with this problem is to develop estimators which are based upon alternative distributional assumptions, and in later chapters we shall adopt this approach when dealing with discrete and ordered response data. Seltzer (1993) gave an example, using Gibbs sampling, based on the assumption that the response variable has a t-distribution, and this approach can be extended to other continuous but skew distributions.

An alternative procedure is to modify the standard error and confidence interval estimates so that they are less dependent on distributional assumptions, of whatever kind. One of the penalties of this is that the resulting significance tests and confidence intervals will tend to be wider, or more 'conservative', than those derived under a particular distributional assumption.

Consider first the fixed part of the model and the usual IGLS estimate of the fixed parameters based upon the random parameter estimates

$$\hat{\beta} = (X^T \hat{V}^{-1} X)^{-1} X^T \hat{V}^{-1} Y$$

The covariance matrix of these estimates is

$$\text{cov}(\hat{\beta}) = (X^T \hat{V}^{-1} X)^{-1} X^T \hat{V}^{-1} \{\text{cov}(Y)\} \hat{V}^{-1} X (X^T \hat{V}^{-1} X)^{-1} \tag{3.8}$$

where $\text{cov}(Y) = V$ and is unknown. The usual procedure is to substitute the estimated $\hat{V}$, but this will generally lead to standard errors which are too small. A robust estimator is obtained by replacing $\text{cov}(Y)$ by $\tilde{Y}\tilde{Y}^T$, namely the cross product matrix of the raw residuals, which is a consistent estimator of V. This is done for each highest level block of V in order to satisfy the block diagonality structure of the model. This estimator is a generalization of the estimator given by Royall (1986) for a single level model which uses only the diagonal elements of $\tilde{Y}\tilde{Y}^T$.

For the random parameters, an analogous result holds. It is also possible to derive robust estimators for residuals, but these generally are not useful because the estimate for each residual corresponding to a higher level unit uses the corresponding value of $\tilde{Y}\tilde{Y}^T$ and this can give very unstable estimates.

We now apply (3.8) to the abortion data analyses, and Table 3.5 shows the result for analysis A of Table 3.4 and an OLS analysis. The major change is in the estimate of the standard error for level 1, with only moderate changes for the fixed parameters.

Another approach to providing robust standard errors is to use jackknifing (Miller, 1974). Thus, if we wished to calculate the standard error for a level 2 variance in a model with p level 2 units, the jackknife procedure would involve recomputing the variance for p subsamples, each one formed by omitting one level 2 unit, and using the set of these to form the standard error estimate. The procedure

Table 3.5 Robust standard errors for analysis A in Table 3.4

Parameter	Estimate	Model based s.e.	Robust s.e.
Fixed:			
Constant	0.32		
Religion: R. Catholic	−0.80	0.176	0.225
Protestant	−0.27	0.098	0.102
Other	−0.63	0.127	0.121
Year: 1984	−0.29	0.048	0.050
1985	−0.06	0.048	0.061
1986	0.06	0.048	0.047
Random:			
Level 3			
σ_v^2	0.03	0.030	0.028
Level 2			
σ_u^2	0.37	0.043	0.055
Level 1			
σ_{e0}^2	0.31	0.016	0.031

also gives a revised estimate of the parameter itself. Longford (1993, Chapter 6) gives an example in the analysis of a complex matrix sample design and suggests that there may be often a considerable loss of efficiency using the jackknife method, and it is also computationally intensive.

A more flexible method is that of bootstrapping (see Efron and Gong, 1983 for an introduction; and Laird and Louis, 1987, 1989 for more extensive discussions in the context of a multilevel model). The basic non-parametric bootstrap procedure involves simple random resampling with replacement of the response variable values (or residuals in a linear model) to generate a single bootstrap sample. The model parameter estimates are then re-estimated for this sample. This procedure is repeated a large number (N) of times yielding N sets of parameter estimates, which are then treated as a simple random sample and used to derive standard errors or confidence intervals. For a multilevel model, however, such a procedure is inadequate since it assumes identically distributed responses, although for certain models it may be possible to adapt this procedure (see for example Moulton and Zeger, 1989).

The fully parametric bootstrap utilizes the distributional assumptions of the model in order to generate simulated values which are used to estimate bootstrap sets of parameters. Consider the simple 2-level model assuming Normality

$$y_{ij} = (X\beta)_{ij} + u_j + e_{ij}, \quad \text{var}(u_j) = \sigma_u^2, \quad \text{var}(e_{ij}) = \sigma_e^2$$

To generate a bootstrap sample we select at random from $N(0, \sigma_u^2)$ a set of level 2 values u_j^* and for each level 2 unit a set of e_{ij}^* from $N(0, \sigma_e^2)$. These are added to $(X\beta)_{ij}$ to generate a set of pseudo-values y_{ij}^*, which is then treated as a set of responses from which a new set of bootstrap parameter values, $\hat{\beta}^*, \hat{\sigma}_u^{*2}, \hat{\sigma}_e^{*2}$ is obtained.

Once the set of bootstrap values is available we can use these values to estimate the parameter covariance matrices or standard errors using the usual sample proce-dures. Confidence intervals for the original parameter estimates, or functions of them, can be constructed from these by assuming Normality. Alternatively, we can construct intervals non-parametrically from the percentiles of the set of empirical bootstrap values and where the median value for a parameter or function of parameters deviates substantially from the original parameter estimate, a bias correction procedure should be used. This involves smoothing the bootstrap distribution using a standard Normal distribution. We first estimate z_0, which is the standard Normal score corresponding to the percentile position of the original parameter estimate. Writing $z^{(1-\alpha)}$, $z^{(\alpha)}$ for the standard Normal deviates corresponding to the required (symmetric) percentiles (for example 5% and 95%), we transform back to the bootstrap distribution from the standard Normal distribution values

$$2z_0 + z^{(1-\alpha)}, \quad 2z_0 + z^{(\alpha)}$$

Efron (1988) discusses this and a further correction based on skewness to improve accuracy.

If we wish to obtain bootstrap estimates for estimated level 2 residuals, then for each bootstrap sample we also estimate the residuals, $\hat{u}_j^*$. To estimate the 'comparative' variance of the residuals for each level 2 unit we need to work with $\tilde{u}_j^* = \hat{u}_j^* - u_j^*$ and then use these directly to estimate the required variance, or covari-ance matrix where there are several random coefficients. They can also be used to construct non-parametric confidence intervals as above.

The parametric bootstrap procedure can be extended straightforwardly to nonlin-ear models as discussed in Chapter 5, and especially to the discrete response mod-els of Chapter 7. The only difference is that with, say, a binary response model, we generate binary (0,1) random variables to produce the pseudo-responses rather than Normally distributed ones. Waclawiw and Liang (1994) gave an example of this using the GEE procedure for obtaining parameter estimates.

Table 3.6 gives parametric bootstrap estimates of standard errors and a central 90% confidence interval based upon a Normality assumption, and also a non-para-metric estimation from 500 bootstrap samples for the model of Table 3.5.

The bootstrap standard errors agree quite well with the model-based ones, except for the level 3 variance. This parameter is based upon only 54 level 3 units as op-posed to 264 level 2 and 1056 level 1 units. This is also reflected in the bootstrap confidence intervals where the non-parametric intervals are fairly close to the Normal theory ones except for the level 3 variance. In general, despite the compu-tational overhead, bootstrap intervals will be desirable where effective sample sizes are small, especially for the random parameters. Where distributions are markedly non-Normal the non-parametric intervals are to be preferred, although these will re-quire considerably more bootstrap samples (typically more than the 500 used here) than are necessary to estimate standard variances and covariances of the bootstrap distribution, where 100 will often suffice.

Aggregate level analyses

3.6   As we discussed in Section 1.12, there are sometimes occasions when the only data available for analysis have already been aggregated to a higher level. For

Table 3.6 Bootstrap standard errors and 90% confidence intervals for Analysis A in Table 3.4

Parameter	Model based s.e.	Bootstrap s.e.	Normal C.I.	Non-parametric Adjusted C.I.
Fixed:				
Religion: R. Catholic	0.176	0.173	(−1.084, −0.516)	(−1.128, −0.532)
Protestant	0.098	0.100	(−0.429, −0.101)	(−0.420, −0.106)
Other	0.127	0.132	(−0.846, −0.414)	(−0.805, −0.377)
Year: 1984	0.048	0.048	(−0.365, −0.209)	(−0.374, −0.216)
1985	0.048	0.047	(−0.140, 0.014)	(−0.141, 0.012)
1986	0.048	0.048	(−0.015, 0.141)	(−0.019, 0.141)
Random:				
Level 3 σ_v^2	0.030	0.022	([0], 0.066)	(0, 0.080)
Level 2 σ_u^2	0.043	0.041	(0.302, 0.436)	(0.308, 0.438)
Level 1 σ_{e0}^2	0.016	0.015	(0.284, 0.334)	(0.288, 0.336)

example, we may have information on student achievement only in terms of the mean achievement for each school, or information on utilization of health services only in terms of the total number of episodes for each administrative area. We examine the possibilities for carrying out analyses with aggregate level data and explore how far these can provide information about the parameters of a more disaggregated model.

Consider the simple model used in Chapter 2 for the Junior School Project data with a response mathematics test score and the earlier mathematics score as a co-variate

$$y_{ij} = \beta_0 + \beta_1 x_{ij} + u_j + e_{ij} \tag{3.9}$$

Suppose that we now aggregate to the school level by averaging over all pupils in each school to obtain

$$y_{.j} = \beta_0 + \beta_1 x_{.j} + u_j + e_{.j} \tag{3.10}$$

If we treat this as a single level model, then the level 1 variance is $\sigma_u^2 + n_j^{-1}\sigma_e^2$ and we can fit the model by specifying two explanatory variables for the random part, namely

$$z_0 = 1, \quad z_{1j} = n_j^{-0.5}$$

with random coefficients e_{0j}, e_{1j} having variances and zero covariance. In many surveys the same number of level 1 units will be sampled from each level 2 unit, in which case a single explanatory variable z_0 will suffice. The main problem with such an analysis is that the estimates will be inefficient compared with those from a

Table 3.7 School level analysis of JSP data

Parameter	Estimate (s.e.) A	Estimate (s.e.) B	Estimate (s.e.) C
Fixed:			
Constant	0.18	0.16	0.16
8-year score	0.091 (0.019)	0.092 (0.020)	0.094 (0.021)
Gender (Propn. boys)	−0.34 (0.30)	−0.31 (0.30)	−0.29 (0.29)
S. Class (Propn. N.M.)	0.00 (0.20)	0.00 (0.28)	−0.01 (0.27)
Random:			
σ_{u0}^2	0.11 (0.021)	0.11 (0.040)	0.08 (0.024)
σ_{e0}^2		0.08 (0.37)	–
σ_{u01}			0.00 (0.01)
σ_{u1}^2			0.004 (0.004)
−2(log-likelihood)	31.33	31.28	29.44

2-level model based on individual student data. Analysis A in Table 3.6 gives the results of an analysis using just the single explanatory variable z_0, and analysis B additionally uses z_{1j} and so is equivalent to a single level weighted regression model. In both analyses we have included the proportion of non-manual students and the proportion of girls as explanatory variables; that is, the average values of the corresponding (0.1) dummy variables.

In comparison with analysis C in Table 3.1, while the coefficient of the 8-year maths score remains unchanged, the coefficients for gender and social class change markedly. We also see how the standard errors are substantially greater. In fact, although the number of students per school varies between 3 and 49, the inclusion of z_{1j} has little effect.

For these data we know that the slope of the 8-year score is random across schools. In this case, model (3.10) becomes

$$y_{.j} = \beta_0 + \beta_1 x_{.j} + u_{0j} + u_{1j} x_{.j} + e_{.j} \tag{3.11}$$

and we obtain the additional contributions to the variance of the aggregated level 2 units

$$\sigma_{u1}^2 x_{.j}^2, \quad 2\sigma_{u01} x_{.j}$$

Analysis C in Table 3.6 shows the results of fitting this model. This is directly comparable with analysis C in Table 3.1 and we can see that although the estimate of the level 2 variance is similar, we have a poor estimate of the random coefficient variance, and, unlike analysis B, it is not possible to estimate a separate level 1 variance because of the small number of units in the analysis.

If there is complex variation at level 1, such as we fitted in Table 3.2, then for such an explanatory variable, say z_{2ij}, we would obtain further contributions to the variance for the aggregated model for unit j

$$2\sigma_{e02} z_{2.j} / n_j, \quad \sigma_{e2}^2 \sum_i z_{2ij}^2 / n_j^2$$

The first of these terms can be fitted as a covariance and the second as a variance, by defining appropriate explanatory variables. In the present case the data are not extensive enough to allow us to fit these additional variables. We also note that the values of the squared explanatory variables in the second of these expressions will often not be available for aggregated data.

If we have an initial 3-level model, and data are aggregated to level 2, we need to specify properly the level 2 random variation resulting from the aggregation process. Failure to do this may allow us to fit random variation at level 3, but any interpretation of this may be problematic because it may have arisen solely as a result of misspecifying the variation at level 2. For example, if we have an explanatory variable which is strongly correlated with the size of the level 2 units, and we fail to include a random coefficient for z_{1j} at level 2, we may well be able to fit a random coefficient for it at level 3, but the usual interpretation of such a coefficient would be inadmissible.

We now look at what happens to the fixed part coefficients when aggregation takes place and we have already seen that the values of the coefficients for gender and social class change. Consider the model

$$y_{ij} = \beta_0 + \beta_1 x_{ij} + \beta_2 x_{.j} + u_j + e_{ij}$$ (3.12)

where the coefficient for $x_{.j}$ in the aggregated model is now $\beta_1 + \beta_2$. We saw in Table 3.1 that the coefficient for the school mean 8-year score was very small, so that we would expect the coefficient for this in the aggregated model to be similar, which Table 3.6 confirms. For gender and social class the coefficients of the corresponding aggregated variables from a 2-level analysis are respectively –0.06 and –0.09 which, when added to the (non-aggregated) coefficients for gender and social class, give values of –0.09 and –0.06 respectively. These are rather different from those in Table 3.6, but the standard errors are very large. Where there is a contextual or compositional effect, whether through the mean aggregated value, or some other statistic derived from the student level distribution as discussed in Section 2.9, then an aggregated analysis will not allow us to obtain separate estimates for the individual and compositional coefficients.

In summary, we have seen that it is sometimes possible to model aggregated data, but this has to be carried out with care, and any interpretations will be constrained by the nature of the true, underlying, non-aggregated model. In addition, the precisions of the estimates obtained from an aggregated analysis will generally be much lower than those obtained from a full multilevel analysis. A discussion of the aggregation issue can also be found in Aitkin and Longford (1986).

Meta analysis

3.7   The term *Meta analysis* (Hedges and Olkin, 1985) refers to the pooling of results of separate studies, all of which are concerned with the same research hypothesis. The aim is to achieve greater accuracy than that obtainable from a single study and also to allow the investigation of factors responsible for between-study variation. Each study typically provides an estimate for an 'effect', for example a group difference, as a 'common' response and the original data are unavailable for analysis. In general, the response measures used will vary, and care is needed in interpreting them as meaning the same thing. Furthermore, the scales of measurement will differ, so that the effect is usually standardized using a suitable within-study estimate of between-unit standard deviation. If the study result derives from a multilevel model, then this estimate will be based on the level 1 variance or, where this is complex, on an estimate pooled over the effect groups being compared. It is important that comparable estimates are used from each study. This implies that the specification of the level 1 units is comparable and that the sources of higher level variation are properly identified. For example, where each study compares teaching methods using a number of schools, the within-school between-student variation would be appropriate for standardization, which implies that the studies concerned should provide estimates of this using suitable multilevel techniques. We consider the case where only a single effect is of interest, but the generalization to the multivariate case is straightforward (see Chapter 4).

For the jth study we define the standardized effect d_j where this is a dimensionless quantity. It may, for example, be a correlation coefficient, a standardized regression coefficient, a group difference, or a weighted group difference. We require an estimate of the variance of d_j, say σ_j^2, and more generally we require the variance of a dimensionless function having the general form

$$\sum_h w_{hj} \hat{\beta}_{hj} / \hat{\sigma}_{ej} \tag{3.13}$$

where the $\hat{\beta}_{hj}$ are parameter estimates from the jth study. For moderately large numbers of level 1 units, we can ignore the variation in the estimate of the level 1 standard deviation ($\hat{\sigma}_{ej}$) and calculate the variance of the numerator of (3.13) using the estimated covariance matrix of the coefficients. Where the number of level 1 units is small, however, we will need to take into account the sampling variance of this estimate and, assuming independence, obtain the required variance using the standard formula for the variance of a ratio of random variables. Hedges and Olkin (1985) discussed a number of procedures for providing such estimates in the single level case. We can now write a simple model as follows

$$d_j = \delta + v_j + u_j, \quad \text{var}(u_j) = \sigma_j^2, \quad \text{var}(v_j) = \sigma_v^2 \tag{3.14}$$

where σ_j^2 is now assumed known and is treated as an offset in the random part of the model (see also Appendix 5.1), δ is the population parameter of interest and σ_v^2 is the between-study variance of the standardized effect. We can add covariates representing study factors to (3.14) in an attempt to explain between-study differences, which is a further aim of Meta Analysis studies. Bryk and Raudenbush (1992) presented an analysis which compared studies of teacher expectations of student ability and attempted to explain study differences.

There are a number of practical problems with Meta Analysis studies. One of these is where the sample of studies used is subject to systematic bias. This can occur, for example, if some studies do not provide sufficient data to estimate a standardized difference and they are a special group. Another common problem arises where the analysis is based upon published studies, and those studies which found 'non-statistically significant' results tend to remain unpublished. This implies that the distribution of results is censored with the smaller ones tending to be missing, a situation known as the publication bias effect. Vevea (1994) discussed the possibility of weighting the studies, that is the units in the model (3.14), using a suitable function of the statistical significance level associated with each effect, in order to compensate for the selective exclusion. Thus, we could carry out a weighted analysis (Section 3.4) where the weights are, say, proportional to the significance level. Vevea also considered the possibility of estimating the weights.

Note

1. In 1984 the seven questions making up the attitude scale were put to respondents in the reverse order, that is with the most 'acceptable' reasons for having an abortion (e.g. as a result of rape) coming first. This illustrates an important issue in surveys of all kinds which collect data for comparisons over time, namely to maintain the same questioning procedure.

4

The Multivariate Multilevel Model

Multivariate multilevel models

4.1 In Chapters 2 and 3 we have considered only a single response variable. We now look at models where we wish simultaneously to model several responses as functions of explanatory variables. As we shall see, the ability to do this provides us with tools for tackling a very wide range of problems. These problems include missing data, rotation or matrix designs for surveys and prediction models. We develop the models using a dataset of examination results.

The data consist of scores on two components of a science examination taken in 1989 by 1905 students in 73 schools and colleges. The examination is the General Certificate of Secondary Education (GCSE) taken at the end of compulsory schooling, normally when students are 16 years of age. The first component is a traditional written question paper (marked out of a total score of 160) and the second consists of coursework (marked out of a total score of 108), including projects undertaken during the course and marked by each student's own teacher. The overall teachers' marks are subject to external 'moderation' using a sample of coursework. Interest in these data centres on the relationship between the component marks at both the school and student level, whether there are gender differences in this relationship and whether the variability differs for the two components. Creswell (1991) has a full description of the dataset.

The basic 2-level multivariate model

4.2 To define a multivariate—in the case of our example, a 2-variate—model we treat the individual student as a level 2 unit and the 'within-student' measurements as level 1 units. Each level 1 measurement 'record' has a response, which is either the written paper score or the coursework score. The basic explanatory variables are a set of dummy variables that indicate which response variable is present. Further explanatory variables are defined by multiplying these dummy variables by

Table 4.1 Data matrix for examination example

Student	Response	Intercepts		Gender	
		Written	Coursework	Written	Coursework
1 (female)	y_{11}	1	0	1	0
1	y_{12}	0	1	0	1
2 (male)	y_{21}	1	0	0	0
2	y_{22}	0	1	0	0
3 (female)	y_{31}	1	0	1	0

individual level explanatory variables, for example gender. The data matrix for three individuals, two of whom have both measurements and the third of whom has only the written paper score, is displayed in Table 4.1. The first and third students are female (1) and the second is male (0).

The model is written as

$$y_{ij} = \beta_{01}z_{1ij} + \beta_{02}z_{2ij} + \beta_{11}z_{1ij}x_j + \beta_{12}z_{2ij}x_j + u_{1j} + u_{2j}$$

$$z_{1ij} = \begin{cases} 1 \text{ if written} \\ 0 \text{ if coursework} \end{cases}, \quad z_{2ij} = 1 - z_{1ij}, \quad x_j = \begin{cases} 1 \text{ if female} \\ 0 \text{ if male} \end{cases} \quad (4.1)$$

$$\text{var}(u_{1j}) = \sigma_{u1}^2, \quad \text{var}(u_{2j}) = \sigma_{u2}^2, \quad \text{cov}(u_{1j}u_{2j}) = \sigma_{u12}$$

There are several features of this model. There is no level 1 variation specified because level 1 exists solely to define the multivariate structure. The level 2 variances and covariance are the (residual) between-student variances. In the case where only the intercept dummy variables are fitted, and since every student has both scores, the model estimates of these parameters become the usual between-student estimates of the variances and covariance. The multilevel estimates are statistically efficient even where some responses are missing and, in the case where the measurements have a multivariate Normal distribution, they are maximum likelihood. Thus, the formulation as a 2-level model allows for the efficient estimation of a covariance matrix with missing responses.

In our example the students are grouped within examination centres, so that the centre is the level 3 unit. Table 4.2 presents the results of two models fitted to these data.

The first analysis is simply (4.1) with variances and a covariance for the two components added at level 3. In the second analysis, additional variance terms for gender have been added.

In both analyses the females do worse on the written paper and better on the coursework assessment. There is a greater variability of marks on the coursework element, even though this is marked out of a smaller total, and the intra-centre correlations are approximately the same in the first analysis (0.28 and 0.30). This suggests that the 'moderation' process has been successful in maintaining a similar relative between-centre variation for the coursework marks. The correlation between the two components is 0.50 at the student level and 0.41 at the centre level.

Table 4.2 Bivariate models for written paper and coursework responses

	Estimate (s.e.)	Estimate (s.e.)
Fixed:		
Constant: Written	49.5	49.5
Coursework	69.5	69.1
Gender: Written	−2.5 (0.5)	−2.5 (0.5)
Coursework	6.9 (0.7)	7.3 (1.1)
Random:		
Level 3		
σ_{v1}^2	48.9 (9.5)	49.6 (9.5)
σ_{v12}	25.2 (9.1)	35.5 (11.3)
σ_{v2}^2	77.1 (14.8)	106.6 (21.7)
σ_{v14}		−15.9 (7.8)
σ_{v24}		−37.4 (13.2)
σ_{v4}^2		41.5 (11.7)
Level 2		
σ_{u1}^2	124.3 (4.1)	124.2 (4.1)
σ_{u12}	74.6 (3.9)	73.6 (3.9)
σ_{u2}^2	183.2 (6.1)	189.1 (8.6)
σ_{u24}		−12.5 (4.7)
−2 log-likelihood	29718.8	29664.7

The subscripts refer to the following explanatory variables: 1 = writing intercept, 2 = coursework intercept, 3 = writing gender, 4 = coursework gender.

In the second analysis we see that the between-student variance for coursework is smaller for the females (164.0) compared with that for the males (189.1) and for the centres the coursework variance for females is also smaller (73.3) than for males (106.6). There appears to be no difference in the variances for the written paper.

Note how the standard error of the coursework gender coefficient increases with the more precise specification of the coursework variation at both levels. This is another aspect of the effect we saw when fitting a multilevel model as opposed to a single level model.

Rotation designs

4.3   We have already seen that fully balanced multivariate designs are unnecessary and randomly missing responses are handled automatically. As Table 4.1 shows, the basic 2-level formulation does not formally recognize that a response is missing, since we only record those present. We now look at designs where responses are effectively missing by design and we see how this can be useful in a number of circumstances.

In many kinds of surveys the amount of information required from respondents is so large that it is too onerous to expect each one to respond to all the questions or items. In education we may require achievement information covering a large number of areas, in surveys of businesses we may wish to have a large amount of detailed information, and in household questionnaires we may wish to obtain information on a wide range of topics. We consider only measurements that are used as responses in a model. If we denote the total set of responses as $\{N\}$ then we choose p subsets $\{N_i, i = 1, \ldots p\}$ each of which is suitable for administering to a subject (level 1).

When choosing these subsets we can only estimate subject-level covariances between those responses that appear together in a subtest. It is therefore common in such designs to ensure that every possible pair of responses is present. If we wish to estimate covariances for higher level units, such as schools, it is necessary only to ensure that the relevant pair of responses are assigned to the same schools—a large enough number to provide efficient estimates. The subjects are assigned at random to subtest, and higher level units are also assigned randomly, possibly with stratification.

Each subset is viewed formally as a multivariate response vector with randomly missing values, although the missing observations are produced by design. As we saw in the previous section, we can fit a multivariate response model for such data and obtain efficient estimates for the fixed part coefficients and covariance structures at any level. In this formulation, the variables to be used as explanatory variables should be measured for each level 1 unit. We shall discuss how to deal with missing explanatory variable values in Chapter 11. We now give an example using educational achievement data.

A rotation design example using science test scores

4.4   The data come from the Second International Science Survey carried out by the International Association for the Evaluation of Educational Achievement (Rosier, 1987). Table 4.3 shows how items from three science topic areas are distributed over test papers or forms, and the numbers of items in each topic area. The tests consisted of a core form taken by all students plus a randomly selected pair out of the four additional forms. The study was carried out in 1984 in some 24 countries. We discuss here the results for Hungary.

Because the number of items in the first additional form was very small, and likewise in some of the other forms for some subjects, only the subsets shown from additional forms 2–4 are used. We also divide each subtest score by the total

Table 4.3 Numbers of items in topic areas: Grade 8

Form	Earth Science	Biology	Physics
1 (Core)	6	10	10
2	–	–	7
3	–	4	–
4	–	4	–

number of items in the subtest so as to reduce each score to the same scale. There are 99 schools with 2439 students and a total of 10971 responses.

We see that the intercorrelations at the student level are low, and are higher at the school level. One reason for this is the fact that there are few items in each subtest so that the reliability of the tests is rather low. This will decrease the correlations at the student level but less so at the school level. In Chapter 10 we shall see how we can make corrections for unreliability. Because of the low reliabilities the joint analysis does not result in a marked improvement in efficiency when we compare this analysis with an analysis for a single subtest. For example, if

Table 4.4 Science attainment estimates for Hungary IEA study

	Estimate (s.e.)
Fixed:	
Earth Science Core	0.838 (0.0076)
Biology Core	0.711 (0.0100)
Biology R3	0.684 (0.0109)
Biology R4	0.591 (0.0167)
Physics Core	0.752 (0.0128)
Physics R2	0.664 (0.0128)
Earth Science Core (girls–boys)	−0.0030 (0.0059)
Biology Core (girls–boys)	−0.0151 (0.0066)
Biology R3 (girls–boys)	0.0040 (0.0125)
Biology R4 (girls–boys)	−0.0492 (0.0137)
Physics Core (girls–boys)	−0.0696 (0.0073)
Physics R2 (girls–boys)	−0.0696 (0.0116)

Random: Variances on diagonal; correlations off-diagonal
Level 2 (School)

	E.Sc. core	Biol. core	Biol. R3	Biol. R4	Phys. core	Phys. R2
E.Sc. core	0.0041					
Biol. core	0.68	0.0076				
Biol. R3	0.51	0.68	0.0037			
Biol. R4	0.46	0.68	0.45	0.0183		
Phys. core	0.57	0.90	0.76	0.63	0.0104	
Phys. R2	0.54	0.78	0.57	0.65	0.78	0.0095

Level 1 (Student)

	E.Sc. core	Biol. core	Biol. R3	Biol. R4	Phys. core	Phys. R2
E.Sc. core	0.0206					
Biol. core	0.27	0.0261				
Biol. R3	0.12	0.13	0.0478			
Biol. R4	0.14	0.27	0.20	0.0585		
Phys. core	0.26	0.42	0.11	0.27	0.0314	
Phys. R2	0.22	0.33	0.14	0.37	0.41	0.0449

we fit a univariate model for the Physics R2 subtest, using the 1226 students responding to that subtest, we obtain fixed part estimates of 0.665 (0.0132) and −0.073 (0.0124) which are close to those above and with standard errors only slightly higher.

In order to provide the most precise estimates we treated the subtests separately, although we would generally wish to make inferences for each subject area, combining over the tests. The natural way to do this is to form a weighted average of the subtest scores, in this case weighting by the number of items in each subtest. Thus, for the biology core and subtests we would form the weighted sum with weights 0.556, 0.222 and 0.222 respectively. This gives estimates for the boys and (girls–boys) of 0.68 (0.009) and −0.02 (0.007). We can compare this with the weighted combination of the core and two subtests, eliminating any students with missing data. This results in only 399 students with complete data and the corresponding estimates are 0.68 (0.013) and −0.008 (0.015). In this case, even though the individual level 1 correlations are relatively small, the gain in efficiency is substantial, especially for inferences about the gender difference, which in the second analysis is less than its standard error.

Another way to combine the subtests would be to form, for each student, a score based upon the items which the student responded to. Thus, for Biology the 399 students taking the core and both rotated forms would have a score out of 18 items; and there would be 823 and 807 students respectively with scores out of 14 items with 410 students having only a score out of the core test. Since the scores are out of different totals, we would expect the between-student and between-school variances to differ and this is the case; the between-student variance for the 10 core test score is 0.00013 compared with that for the 18 item core and two rotated forms score of 0.00021. Thus, we would need to fit separate variance and covariance terms in general for each of the combinations and, in effect, treat the four combinations as separate responses in order to obtain efficient estimates. Furthermore, we would also tend to obtain high correlations between these combination scores that could lead to numerical estimation problems, so that, in general, this procedure is not recommended.

Principal components analysis

4.5   We have already seen in Section 4.1 that the covariance matrix for a multivariate response vector where there are missing data can be efficiently estimated by arranging for the multivariate structure to constitute a 'dummy' level 1. When the variables have a multivariate distribution the resulting estimates are maximum likelihood or restricted maximum likelihood.

The aim of principal components analysis is to find a linear function of a set of variates which has the maximum variance, subject to a suitable constraint. In the single level case we require to maximize the variance of $\mathbf{w}^T\mathbf{y}$ where $\mathbf{w}$ is the vector of weights defining the linear function of the variates $\mathbf{y}$, and Ω is the covariance matrix of $\mathbf{y}$, namely

$$\Lambda = \mathbf{w}^T\Omega\mathbf{w}, \quad \mathbf{w}^T\mathbf{w} = 1$$

The solution is given by the eigenvector associated with the largest eigenvalue of Ω; that is, the solution of

$$|\Omega - \lambda I| = 0 \qquad (4.2)$$

We define a second function by the set of weights that maximizes the variance subject to the function being uncorrelated with the first function. The solution is given by the eigenvalue associated with the second largest eigenvalue, and subsequent functions can be defined similarly (Lawley and Maxwell, 1971). The variates are usually standardized to have equal variances.

We note that the covariance (or correlation) matrix Ω can be a residual matrix, after regressing on explanatory variables. Thus, if we wished to form a principal component for the four science subjects of the previous section, we may wish to use the residual covariance matrix, after adjusting for gender differences. We now, however, have a choice of two covariance matrices, the between-student and the between-school one. If we choose the between-student matrix, then we would interpret the principal component as that which had been adjusted for school differences. In forming the derived summary variable(s) we would not use the actual observed variates but the level 1 estimates of them; that is, the level 1 residuals, the $\hat{u}_{01j}$, $\hat{u}_{02j}$ of (4.1).

We could also choose to summarize the level 2 covariance matrix, and in this case we would use the school level residuals as the variates in the linear function. If the principal component analysis has been carried out on the residuals from a multivariate multilevel analysis then we may wish to regard the school level principal component as a convenient summary measure of school differences.

Table 4.5 shows the student level and school level principal component weights for the Science data. Since the measures are designed to be on the same scale we work directly with the covariance matrices.

As might be expected, the components both have positive weights. At the school level, the percentage variation accounted for by the first component is high, suggesting that school Science performance may usefully be summarized by this weighted function of the individual school level subject residuals. Also, the two sets of weights are fairly similar. This suggests that if we wished to summarize the individual subject scores into a single index, we could do this using the student level weights, or even the weights obtained using the total covariance matrix.

Table 4.5 Principal component weights for science test scores and percentage variation accounted for

Subject	Between-student	Between-school
Earth Science Core	0.17	0.21
Biology Core	0.29	0.40
Biology R3	0.31	0.21
Biology R4	0.63	0.59
Physics Core	0.35	0.46
Physics R2	0.52	0.43
% variation	41%	72%

Multiple discriminant analysis

4.6   Given a set of variates we can seek a linear function of these variates that best discriminates among groups and this leads to the following definition. If $\bar{\mathbf{y}}$ is the vector of group means then we require a set of weights $\mathbf{w}$ such that $\mathbf{w}^T\bar{\mathbf{y}}$ has maximum variance, subject to the within-group variance of $\mathbf{w}^T\mathbf{y}$ being constrained, for example, equal to 1.0. The solution is the vector associated with the largest root of

$$\left|\Omega_B - \lambda\Omega_W\right| = 0$$

for the between-group (Ω_B) and within-group (Ω_W) covariance matrices. For just two groups this gives the usual 'Fisher' discriminant function. As in principal components analysis we can find further vectors that discriminate best, subject to being uncorrelated with all the previous vectors. The function of the variates $\mathbf{w}^T\mathbf{y}$ can then be used, for example, to classify a new unit into the 'nearest' group.

In the 2-level case our groups are the level 2 units so that we require the covariance matrices from both levels. Using the Science data example the first vector is given by the weights 0.41 −0.07 1.00 0.26 0.31 0.13 and accounts for about 48% of the variation. The next two vectors account for 19% and 13%. It is difficult to interpret these weights and the function would seem to have limited usefulness for discriminating between schools.

Other procedures

4.7   There are other applications of multivariate models and we will be using many of the results of this chapter later. We shall also see in Chapters 5 and 7 how mixtures of continuous and discrete response variables can be handled using extensions to the procedures of this chapter. The ability to model bivariate responses is used in Chapter 9 to deal with event duration models.

5

Nonlinear Multilevel Models

Nonlinear models

5.1 The models of Chapters 1–4 are linear in the sense that the response is a linear function of the parameters in the fixed part and the elements of V are linear functions of the parameters in the random part. In many applications, however, it is appropriate to consider models where the fixed or random parts of the model, or both, contain nonlinear functions. For example, in the study of growth, Jenss and Bayley (1937) proposed the following function to describe the growth in height of young children

$$y_{ij} = \alpha_0 + \alpha_1 t_{ij} + u_{\alpha 0j} + u_{\alpha 1j} t_{ij} + e_{\alpha ij} - \exp(\beta_0 + \beta_1 t_{ij} + u_{\beta 0j} + u_{\beta 1j} t_{ij} + e_{\beta ij}) \quad (5.1)$$

where t_{ij} is the age of the jth child at the ith measurement occasion. Generalized linear models (McCullagh and Nelder, 1989) are a special case of nonlinear models where the response is a nonlinear function of a fixed part linear predictor. Models for discrete data, such as counts or proportions, fall into this category and we shall devote Chapter 7 to studying these. For example, a 2-level log-linear model can be written

$$E(m_{ij}) = \pi_{ij}, \quad \pi_{ij} = \exp(X_{ij}\beta_j) \quad (5.2)$$

where m_{ij} is assumed typically to have a Poisson distribution, in this case across level 1 units. Note here that in the multilevel extension of the standard single level model, the linear predictor contains random variables defined at level 2 or above.

In this chapter we consider a general nonlinear model. Later chapters will use the results for particular applications.

Nonlinear functions of linear components

5.2 The following results are an extension of those presented by Goldstein (1991) and Appendix 5.1 gives details. Where the random variables are not part of the nonlinear function, the procedure gives maximum likelihood estimates (see Appendix 5.1). In the case where the level 1 variation is non-Normal the procedure can be regarded as a generalization of quasi-likelihood estimation (McCullagh and Nelder, 1989) and such models are discussed in Chapter 7.

Restricting attention to a 2-level structure we can write a fairly general model as follows

$$y_{ij} = X_{1ij}\beta_1 + Z_{1ij}^{(2)}u_{1j} + Z_{1ij}^{(1)}e_{1ij} + f(X_{2ij}\beta_2 + Z_{2ij}^{(2)}u_{2j} + Z_{2ij}^{(1)}e_{2ij}) + ... \qquad (5.3)$$

where the function f is nonlinear and where the $+...$ indicates that additional nonlinear functions can be included, involving further fixed part explanatory variables X or random part explanatory variables at levels 1 and 2, respectively $Z^{(1)}$, $Z^{(2)}$. The model is first linearized by a suitable Taylor series expansion and this leads to consideration of a linear model where the explanatory variables in f are transformed using first and second derivatives of the nonlinear function. Note that the linear component of (5.3) is treated in the standard way, and that the random variables at a given level in the linear and nonlinear components may be correlated.

Consider the nonlinear function f. Appendix 5.1 shows that we can write this as the sum of a fixed part component and a random part. The Taylor expansion for the random part up to a second-order approximation for the ijth unit is as follows

$$f_{ij} = f_{ij}(H_{t+1}) + (Z_{2ij}^{(2)}u_{2j} + Z_{2ij}^{(1)}e_{2ij})f_{ij}'(H_t)$$
$$+ (Z_{2ij}^{(2)}u_{2j} + Z_{2ij}^{(1)}e_{2ij})^2 f_{ij}''(H_t)/2 \qquad (5.4)$$

The first term on the right-hand side is the fixed part value of f at the current $((t+1)$th) iteration of the IGLS or RIGLS algorithm; that is, ignoring the random part. The other two terms involve the first and second differentials of the nonlinear function evaluated at the current values from the previous iteration. We have

$$E(Z_{2ij}^{(2)}u_{2j} + Z_{2ij}^{(1)}e_{2ij}) = 0, \quad E(Z_{2ij}^{(2)}u_{2j} + Z_{2ij}^{(1)}e_{2ij})^2 = \sigma_{zu}^2 + \sigma_{ze}^2$$
$$\sigma_{zu}^2 = Z_{2ij}^{(2)} \Omega_u Z_{2ij}^{(2)T}, \quad \sigma_{ze}^2 = Z_{2ij}^{(1)} \Omega_e Z_{2ij}^{(1)T} \qquad (5.5)$$

We write the expansion for the fixed part value as

$$f_{ij}(H_{t+1}) = f_{ij}(H_t) + X_{ij}(\beta_{1,t+1} - \beta_{1,t})f_{ij}'(H_t) \qquad (5.6)$$

where $\beta_{1,t+1}$, $\beta_{1,t}$ are the current and previous iteration values of the fixed part coefficients.

Either we can choose H_t to be the current value of the fixed part predictor, that is $X_{2ij}\beta_2$, or we can add the current estimated residuals to obtain an improved approximation to the nonlinear component for each unit. The former is referred to as a 'marginal' (quasi-likelihood) model and the latter as a 'penalized' or 'predictive' (quasi-likelihood) model (see Breslow and Clayton, 1993, for a further discussion). We can also choose whether or not to include the term in (5.4) involving the second derivative and we would expect its inclusion, in general, to improve the estimates. Its inclusion defines a further offset for the fixed part and one for the random part (see Appendix 5.1). We shall illustrate the effect of these choices in the examples given in Chapter 7. Further details of the estimation procedure are given in Appendix 5.1.

In practice, general models such as (5.1) may pose considerable estimation

Table 5.1 Differentials for some common nonlinear models

Model	Function $f(x)$	First differential $f'(x)$	Second differential $f''(x)$
loglinear	e^x	e^x	e^x
logit	$(1+e^{-x})^{-1}$	$(1+e^x)^{-1}(1+e^{-x})^{-1}$	$(1+e^{-x})^{-1}(1+e^x)^{-2}(1-e^x)$
log-log	e^{-e^x}	$-e^x e^{-e^x}$	$(e^x-1)e^x e^{-e^x}$
inverse	x^{-1}	$-x^{-2}$	$2x^{-3}$

problems. We notice that the same explanatory variables occur in the linear and nonlinear components and this can lead to instability and failure to converge. Further work in this area is required.

Table 5.1 gives expressions for the first and second differentials for some commonly used nonlinear models.

Estimating population means

5.3   Consider the expected value of the response for a given set of covariate values. Because of the nonlinearity this is not, in general, equal to the predicted value when the random variables in the nonlinear function are zero. For example, if we write the variance components model (5.2)

$$\pi_{ij} = \exp(\beta_0 + \beta_1 x_{ij} + u_j)$$

and assuming Normality for u_j, we obtain

$$E(\pi_{ij} \mid x_{ij}) = \exp(\beta_0 + \beta_1 x_{ij}) \int_{-\infty}^{\infty} e^{u_j} \phi(u_j) du_j = \exp(\beta_0 + \beta_1 x_{ij} + \sigma_u^2 / 2)$$

where ϕ is the density function of the Normal distribution. Zeger *et al* (1988) considered this issue and proposed a 'population average' model for directly obtaining population predicted values by eliminating random variables from the nonlinear component. In general, however, this approach is less efficient when the full model with random variables within the nonlinear function is the correct model. The population predicted values, conditional on covariates, can be obtained if required, as above, by taking expectations over the population. An approximation to this can be obtained from the second-order terms in (5.1.4) with higher order terms introduced if necessary to obtain a better approximation. Alternatively we may generate a large number of simulated sets of values for the random variables and, for each set, evaluate the response function to obtain an estimate of the full population distribution.

Nonlinear functions for variances and covariances

5.4   We saw in Chapter 3 how we could model complex functions of the level 1 variance. As with the linear component of the model, there are cases where we may wish to model variances or covariances as nonlinear functions. In principle we can do this at any level but we restrict our attention to level 1 and to the variance only. In Chapter 6 we give an example where the covariances are modelled in this way.

Suppose that the level 1 variance decreases with increasing values of an explanatory variable such that it approaches a fixed value asymptotically. We could then model this for a 2-level model, say, as follows

$$\text{var}(e_{ij}) = \exp(\beta_0^* - \beta_1^* x_{ij})$$

where β_0^*, β_1^* are parameters to be estimated. Such a model also guarantees that the level 1 variance is positive, which is not the case with linear models, such as those based on polynomials. The estimation procedure is analogous to that described above and details are given in Appendix 5.1.

Examples of nonlinear growth and nonlinear level 1 variance

5.5   We give first an example of a model with a nonlinear function for the linear component and we then consider the case of a nonlinear level 1 variance function.

We use an example from child growth, consisting of 577 repeated measurements of height on 197 French Canadian boys aged from 5 to 10 years (Demirjian *et al*, 1982) with between three and seven measurements each. This is a 2-level structure with measurement occasions nested within children. We fit the following version of the Jenss–Bayley curve to illustrate the procedure

$$y_{ij} = \exp(\beta_0 + \beta_1 t_{ij} + \beta_2 t_{ij}^2 + \beta_3 t_{ij}^3 + u_{\beta 0j} + u_{\beta 1j} t_{ij}) + \alpha_0 + e_{\alpha ij} \tag{5.7}$$

so that the fixed part is an intercept plus a nonlinear component and the random part variance at level 2 is part of the nonlinear component. The results are given in Table 5.2, using the first-order approximation with prediction based upon the fixed part only. We shall compare the performance of the different approximations in Chapter 7.

The level 1 variance is small and of the order of the measurement error of height measurements. The starting values for this model need to be chosen with care, and in the present case the model was run to convergence without the linear intercept α_0 which was then added with a starting value of 100. Bock (1992) used an EM algorithm to fit a nonlinear 2-level model to growth data from age 2 years to adulthood using a mixture of three logistic curves.

The second example uses the JSP dataset where we studied the level 1 variance in Chapter 3. We will fit model B of Table 3.1 with a nonlinear function of the level 1 variance, instead of the level 1 variance as a quadratic function of the 8-year score. This level 1 variance for the *ij*th level 1 unit is $\exp(\beta_0^* + \beta_1^* x_{1ij})$ and Table 5.3 shows the model estimates. The estimates are almost identical to those of model B of Table 3.1 as is the likelihood value.

Figure 5.1 shows the predicted level 1 variance for this model and model B of

Table 5.2 Nonlinear model estimates with first-order fixed part prediction. Age is measured about 8.0 years

Fixed coefficient	Estimate (s.e.)
Intercept (linear)	90.3
Intercept (nonlinear)	3.58
Age	0.15 (0.10)
Age squared	−0.016 (0.02)
Age cubed	0.002 (0.004)

Nonlinear model level 2 covariance matrix (s.e.)

	Intercept	Age
Intercept	0.025 (0.003)	
Age	−0.0027 (0.0003)	0.00036 (0.00005)

Level 1 variance = 0.25

Table 3.1. In these data the nonlinear function gives very similar results to the quadratic one. It is clear, however, that where the variance asymptotically approaches a constant value, for extreme values of an explanatory variable, a linear or even quadratic approximation may be expected to fail. In the present case a linear function does predict a negative level 1 variance within the range of the data. An example where a nonlinear function is necessary is in growth data, described in Chapter 6, where the level 1 (within-individual) variation will decrease towards a constant value at the approach to adulthood.

Table 5.3 Nonlinear level 1 variance for JSP data

Parameter	Estimate (s.e.)
Fixed:	
Constant	31.7
8-year score	0.58 (0.03)
Gender (boys–girls)	−0.34 (0.27)
Social class (Non Man.–Man.)	0.76 (0.30)
School mean 8-year score	0.01 (0.11)
8-yr score × school mean 8-yr score	0.02 (0.01)
Random:	
Level 2	
σ_{u0}^2	2.87 (0.88)
σ_{u01}	−0.17 (0.07)
σ_{u1}^2	0.012 (0.007)
Level 1	
β_0*	2.74 (0.06)
β_1*	−0.10 (0.01)

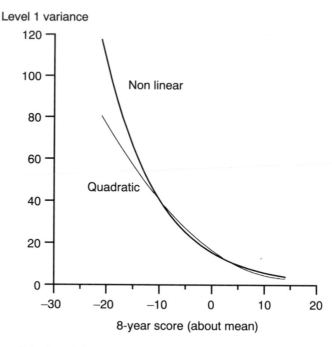

Figure 5.1   Level 1 variance as a function of 8-year Maths score.

Multivariate nonlinear models

5.6   We can use the procedures of this chapter to fit multivariate models by using level 1 to define the multivariate structure and using the linearization procedures described in this chapter for higher levels. In general, the response variables will have different nonlinear link functions, some of which may be linear. Thus, for example, we might fit a model where one response, say a mathematics test score, is a linear function of explanatory variables and a second response, say whether or not the student has a 'positive' attitude towards mathematics, is binary. For each level we will have variances for each response and covariances among the coefficients that are random at that level, where these are specified for the transformed model. Such a model is discussed in Chapter 7.

We may also have multivariate models where the level 1 variances are different nonlinear or linear functions of explanatory variables with covariances between the coefficients in the different nonlinear or linear functions.

APPENDIX 5.1

Nonlinear model estimation

Modelling linear components

5.1.1 We consider first only a single nonlinear term of the form

$$y_{ij} = f(X_{2ij}\beta_2 + Z_{2ij}^{(2)}u_{2j} + Z_{2ij}^{(1)}e_{2ij})$$ (5.1.1)

The addition of linear terms to this model is discussed in Chapter 5.

At the $(t + 1)$th iteration we expand (5.1.1) for both fixed and random parts as follows

$$f_{ij}(H_t) + X_{ij}(\beta_{2,t+1} - \beta_{2,t})f'_{ij}(H_t)$$
$$+ (Z_{2ij}^{(2)}u_{2j} + Z_{2ij}^{(1)}e_{2ij})f'_{ij}(H_t) + (Z_{2ij}^{(2)}u_{2j} + Z_{2ij}^{(1)}e_{2ij})^2 f''_{ij}(H_t)/2$$ (5.1.2)

in terms of parameter values estimated at the tth iteration. The first line of (5.1.2) updates the fixed part of the model and, in the special case of a single level quasi-likelihood model, provides the updating function. The quantity $f_{ij}(H_t) - X_{2ij}\beta_{2,t}f'_{ij}(H_t)$ is treated as an offset to be subtracted from the response variable. The first term in the second line defines a linear random component based on the explanatory variables transformed by multiplying by the first differential. We need to specify H_t and consider the distribution of the second term in the second line of (5.1.2).

If we choose $H_t = X_{2ij}\beta_{2,t}$, this is equivalent to carrying out the Taylor expansion around the fixed part predicted value. If we choose $H_t = X_{2ij}\beta_{2,t} + Z_{2ij}^{(2)}\hat{u}_{2j} + Z_{2ij}^{(1)}\hat{e}_{2ij}$, this expands around the current predicted value for the ijth unit and we replace the second line of (5.1.2) by

$$(Z_{2ij}^{(2)}(u_{2j} - \hat{u}_{2j}) + Z_{2ij}^{(1)}(e_{2ij} - \hat{e}_{2ij}))f'_{ij}(H_t)$$
$$+ (Z_{2ij}^{(2)}(u_{2j} - \hat{u}_{2j}) + Z_{2ij}^{(1)}(e_{2ij} - \hat{e}_{2ij}))^2 f''_{ij}(H_t)/2$$

We thus have the further offset from the linear term to be added to the response

$$(Z_{2ij}^{(2)}\hat{u}_{2j} + Z_{2ij}^{(1)}\hat{e}_{2ij})f'_{ij}(H_t)$$

A discussion of these approaches in the context of multilevel generalized linear models is given by Breslow and Clayton (1993). Wolfinger (1993) synthesizes some of the literature based upon this 'predictive' approach. All these methods use only the first-order terms in (5.1.2).

From the second line of (5.1.2), where the Taylor expansion is about zero, we have

$$E(Z_{2ij}^{(2)}u_{2j} + Z_{2ij}^{(1)}e_{2ij}) = 0, \quad E(Z_{2ij}^{(2)}u_{2j} + Z_{2ij}^{(1)}e_{2ij})^2 = \sigma_{zu}^2 + \sigma_{ze}^2$$

$$\sigma_{zu}^2 = Z_{2ij}^{(2)}\Omega_u Z_{2ij}^{(2)T}, \quad \sigma_{ze}^2 = Z_{2ij}^{(1)}\Omega_e Z_{2ij}^{(1)T} \tag{5.1.3}$$

To incorporate the second-order terms we treat $(\sigma_{zu}^2 + \sigma_{ze}^2)f''(H_t)/2$ as an additional offset in the fixed part and in the random part of the model we need to consider the variation of the second term in the second line of (5.1.2). If we assume Normality then all third moments, formed from the product of the two terms in the second line of (5.1.2), are zero and we have

$$\text{var}(Z_{2ij}^{(2)}u_{2j} + Z_{2ij}^{(1)}e_{2ij})^2 = 2(\sigma_{zu}^4 + \sigma_{ze}^4) \tag{5.1.4}$$

so that we need to define the additional random variables

$$Z_u^* = \sigma_{zu}^2 f''(H_t)/\sqrt{2}, \quad Z_e^* = \sigma_{ze}^2 f''(H_t)/\sqrt{2}$$

which are uncorrelated and with variances constrained to be equal to 1.0. Equivalently we can form $Z_u^* Z_u^{*T}$, $Z_e^* Z_e^{*T}$ as offsets for the response vector $\text{vec}(\tilde{Y}\tilde{Y}^T)$ in the estimation of the random parameters. Having modified the response variable by removing the necessary offsets we are left in the fixed part with a modified response, say Y', with a modified explanatory variable matrix, say X'. We do likewise for the random part of the model and then carry out a standard iterative procedure, updating the differential functions at each iteration.

Where the Taylor expansion is taken about the current values of the residuals we require the value of

$$E[Z_{2ij}^{(2)}(u_{2j} - \hat{u}_{2j})]^2 + E[Z_{2ij}^{(1)}(e_{2ij} - \hat{e}_{2ij})]^2$$

which leads to the 'conditional' variances described in Appendix 2.2, so that we substitute these variances, $\tilde{\Omega}_u$ and $\tilde{\Omega}_e$, for Ω_u and Ω_e in the above expressions for the fixed and random offsets.

To estimate residuals we note that, having adjusted the response using the offsets, we have, on the right-hand side of the model, for the Taylor expansion about zero, the fixed part together with the random terms

$$(Z_{2ij}^{(2)}u_{2j} + Z_{2ij}^{(1)}e_{2ij})f_{ij}'(H_t) + [(Z_{2ij}^{(2)}u_{2j} + Z_{2ij}^{(1)}e_{2ij})^2 - (\sigma_{zu}^2 + \sigma_{ze}^2)]f_{ij}''(H_t)/2$$

Each residual and its square appear in this expression, and since third-order moments are zero, we can apply the usual linear estimation for the residuals as described in Appendix 2.2. The weight matrix V is based upon both the linear and quadratic terms of the above expression. We carry out an analogous procedure for the case where the Taylor expansion is based upon the current residual estimates.

The above can be extended in a straightforward way to more than two levels and, of course, to multivariate models. For the first-order approximation the procedure outlined here is closely related to that given by Lindstrom and Bates (1990), for 2-level repeated measures data, who considered a first-order expansion about the unit-specific predicted values. Gumpertz and Pantula (1992) considered a variance components model where the fixed part predictor is nonlinear.

For generalized linear models, Waclawiw and Liang (1993) considered a generalized estimating equations approach (see Chapter 2), using a unit-specific predictor. A full likelihood based method for a repeated measures model with binary responses is given by Garret *et al* (1993).

For small samples, as discussed in Appendix 2.1, we should use the unbiased (RIGLS, REML) procedure to obtain corresponding unbiased quasi-likelihood estimates.

Modelling variances and covariances as nonlinear functions

5.1.2   In Section 2.6 we saw that the random parameters were estimated by regressing the observed cross-product matrix of residuals on a set of explanatory variables which defined the appropriate variances and covariances at each level. Using the notation in Appendix 2.2, we have the following *linear* model for the random parameters β^*

$$Y^* = \text{vec}(\tilde{Y}\tilde{Y}^T) = X^*\beta^*, \quad E(Y^*) = \text{vec}(V) \qquad (5.1.5)$$

We can now apply the same procedure for the specification and estimation of a nonlinear model as above. We illustrate this for the case where the level 1 variance is an exponential function of a covariate X_1^*, defined in terms of the Kronecker product as in Appendix 2.2; namely, for the *t*th element of $X^*\beta^*$ (which is on the diagonal of V) the level 1 variance contribution is

$$\sigma_{et}^2 = f(\beta^*) = \exp(\beta_0^* x_{0t}^* + \beta_1^* x_{1t}^*), \quad X_1^* = \{x_{1t}^*\}, \quad \beta^* = \begin{pmatrix} \beta_0^* \\ \beta_1^* \end{pmatrix} \qquad (5.1.6)$$

As in the linear function case we form the first differential $f' = f$, multiply x_{0t}^*, x_{1t}^* by this and estimate the parameters of the resulting transformed linear model. This will involve introducing an offset for Y^* and constructing the following level 1 explanatory variables for the estimation of β^*, setting the covariance to zero

$$\{x_{0t}^* \exp(\beta_0^* x_{0t}^* + \beta_1^* x_{1t}^*)\}^{0.5}, \quad \{x_{1t}^* \exp(\beta_0^* x_{0t}^* + \beta_1^* x_{1t}^*)\}^{0.5}$$

Because we are estimating only nonlinear functions of linear components here and not adding approximations to a further random component, the estimates obtained are exact maximum likelihood or restricted maximum likelihood estimates.

In Chapter 5 we give an example of model (5.1.6) and in Chapter 6 we develop a special case of a nonlinear model for covariances. We note that the parameters β_0^*, β_1^* are not necessarily positive when modelling (5.1.6) and although we would normally regard such level 1 parameters as variances, in this case as in Section 3.1, they are simply parameters to be estimated. As with nonlinear modelling, in general it is important to have reasonable starting values. These might be obtained by trial and error or by making preliminary estimates of variances for various values of the relevant explanatory variable and regressing their logarithms on the level 1 explanatory variables.

Likelihood values

5.1.3   The log-likelihood for the general multilevel model, apart from a constant and assuming multivariate Normality, is given (Appendix 2.2) by

$$\log L = -\mathrm{tr}(V^{-1}S) - \log|V|, \quad S = \tilde{Y}\tilde{Y}^{\mathrm{T}}, \quad \tilde{Y} = Y - X\beta \qquad (5.1.7)$$

An approximation to this for nonlinear models of a linear component is given by substituting the nonlinear function $f(X\beta)$ for $X\beta$ in (5.1.7) with the transformed random parts of the model incorporated into V in the usual way. If we use the predicted residuals to form H_t then we omit these from the likelihood calculation but add the offset term defined in (5.1.2) to $X\beta$. Likewise, in the second-order model, we have to add the corresponding offsets to V. This procedure is equivalent to computing the ordinary likelihood using the modified response and explanatory variables Y', X' at convergence.

The estimates of $-2\log L$ computed in this way can be used for approximate tests of hypotheses and for constructing confidence intervals. In Chapter 7, when we consider discrete data models with non-Normal level 1 random variation (for example binomial) we may often be able to treat this variation as approximately Normal and carry out the same procedure, and such a procedure will give us an approximate log quasi-likelihood which may be used similarly. In some cases, however, for example when the responses are binary (0,1) this statistic is too unreliable to use and we can base approximate inferences upon the estimated variances and covariances of the parameters. More accurate inferences based upon bootstrap confidence intervals can be obtained as described in Chapter 3.

When modelling variances and covariances as nonlinear functions the estimates obtained are exact maximum likelihood, as is the value of $-2\log L$.

6

Models for Repeated Measures Data

Models for repeated measures

6.1 When measurements are repeated on the same subjects, for example students or animals, a 2-level hierarchy is established with measurement repetitions or occasions as level 1 units and subjects as level 2 units. Such data are often referred to as 'longitudinal', as opposed to 'cross-sectional', where each subject is measured only once. Thus, we may have repeated measures of body weight on growing animals or children, repeated test scores on students or repeated interviews with survey respondents. It is important to distinguish two classes of models which use repeated measurements on the same subjects. In one, earlier measurements are treated as covariates rather than responses. This was done for the educational data analysed in Chapters 2 and 3, and will often be more appropriate when there are a small number of discrete occasions and where different measures are used at each occasion. In the other, usually referred to as 'repeated measures', models all the measurements are treated as responses, and it is this class of models we shall discuss here. A detailed description of the distinction between the former 'conditional' models and the latter 'unconditional' models can be found in Goldstein (1979) and Plewis (1985).

We may also have repetition at higher levels of a data hierarchy. For example, we may have annual examination data on successive cohorts of 16-year-old students in a sample of schools. In this case the school is the level 3 unit, year is the level 2 unit and student the level 1 unit. We may even have a combination of repetitions at different levels: for example, in the previous example, with the students themselves being measured on successive occasions during the years when they take their examination. We shall also look at an example where there are responses at both level 1 and level 2, that is specific to the occasion and to the subject. It is worth pointing out that in repeated measures models, typically most of the variation is at level 2, so that the proper specification of a multilevel model for the data is of particular importance.

The link with the multivariate data models of Chapter 4 is also apparent where the occasions are fixed. For example, we may have measurements on the height of a sample of children at ages 11.0, 12.0, 13.0 and 14.0 years. We can regard this as having a multivariate response vector of four responses for each child, and perform an equivalent analysis, for example relating the measurements to a polynomial function of age. This multivariate approach has traditionally been used with repeated measures data (Grizzle and Allen, 1969). It cannot, however, deal with data with an arbitrary spacing or number of occasions and we shall not consider it further.

In all the models considered so far we have assumed that the level 1 residuals are uncorrelated. For some kinds of repeated measures data, however, this assumption will not be reasonable, and we shall investigate models which allow a serial correlation structure for these residuals.

We deal only with continuous response variables in this chapter. We shall discuss repeated measures models for discrete response data in Chapter 7.

A 2-level repeated measures model

6.2   Consider a data set consisting of repeated measurements of the heights of a random sample of children. We can write a simple model

$$y_{ij} = \beta_{0j} + \beta_{1j}x_{ij} + e_{ij} \tag{6.1}$$

This model assumes that height (Y) is linearly related to age (X) with each subject having their own intercept and slope so that

$$E(\beta_{0j}) = \beta_0, \quad E(\beta_{1j}) = \beta_1$$

$$\text{var}(\beta_{0j}) = \sigma_{u0}^2, \quad \text{var}(\beta_{1j}) = \sigma_{u1}^2, \quad \text{cov}(\beta_{0j}, \beta_{1j}) = \sigma_{u01}, \quad \text{var}(e_{ij}) = \sigma_e^2$$

There is no restriction on the number or spacing of ages, so we can fit a single model to subjects who may have one or several measurements. We can clearly extend (6.1) to include further explanatory variables, measured either at the occasion level, such as time of year or state of health, or at the subject level such as birthweight or gender. We can also extend the basic linear function in (6.1) to include higher order terms and we can further model the level 1 residual so that the level 1 variance is a function of age.

We explored briefly a nonlinear model for growth measurements in Chapter 5. Such models have an important role in certain kinds of growth modelling, especially where growth approaches an asymptote as in the approach to adult status in animals. In the following sections we shall explore the use of polynomial models, which have a more general applicability and for many applications are more flexible (see Goldstein, 1979 for a further discussion). We introduce examples of increasing complexity, and include some nonlinear models for level 1 variation using the results of Chapter 5.

A polynomial model example for adolescent growth and the prediction of adult height

6.3   Our first example combines the basic 2-level repeated measures model with a multivariate model to show how a general growth prediction model can be constructed.

The data consist of 436 measurements of the heights of 110 boys between the ages of 11 and 16 years together with measurements of their height as adults and estimates of their bone ages at each height measurement based upon wrist radiographs. A detailed description can be found in Goldstein (1989b). We first write down the three basic components of the model, starting with a simple repeated measures model for height using a 5th degree polynomial.

$$y_{ij}^{(1)} = \sum_{h=0}^{5} \beta_h^{(1)} x_{ij}^h + \sum_{h=0}^{2} u_{hj}^{(1)} x_{ij}^h + e_{ij}^{(1)} \tag{6.2}$$

where the level 1 term e_{ij} may have a complex structure, for example a decreasing variance with increasing age.

The measure of bone age is already standardized since the average bone age for boys of a given chronological age is equal to this age for the population. Thus, we model bone age using an overall constant to detect any average departure for this group together with between-individual and within-individual variation.

$$y_{ij}^{(2)} = \beta_0^{(2)} + \sum_{h=0}^{1} u_{hj}^{(2)} x_{ij}^h + e_{ij}^{(2)} \tag{6.3}$$

For adult height we have a simple model with an overall mean and level 2 variation. If we had more than one adult measurement on individuals we would also be able to estimate the level 1 variation among adult height measurements; in effect, the measurement errors.

$$y_j^{(3)} = \beta_0^{(3)} + u_{0j}^{(3)} \tag{6.4}$$

We now combine these into a single model using the following indicators

$$\delta_{ij}^{(1)} = 1, \text{ if growth period measurement,} \quad 0 \text{ otherwise}$$

$$\delta_{ij}^{(2)} = 1, \text{ if bone age measurement,} \quad 0 \text{ otherwise}$$

$$\delta_j^{(3)} = 1, \text{ if adult height measurement,} \quad 0 \text{ otherwise}$$

$$y_{ij} = \delta_{ij}^{(1)} \left(\sum_{h=0}^{4} \beta_h^{(1)} x_{ij}^h + \sum_{h=0}^{2} u_{hj}^{(1)} + e_{ij}^{(1)} \right) + \delta_{ij}^{(2)} \left(\beta_0^{(2)} + \sum_{h=0}^{1} u_{hj}^{(2)} x_{ij}^h + e_{ij}^{(2)} \right)$$
$$+ \delta_j^{(3)} \left(\beta_0^{(3)} + u_{0j}^{(3)} \right) \tag{6.5}$$

At level 1, the simplest model, which we shall assume, is that the residuals for bone age and height are independent, although dependencies could be created, for example if the model was incorrectly specified at level 2. Thus, level 1 variation is specified in terms of two variance terms. Although the model is strictly a multivariate model, because the level 1 random variables are independent it is unnecessary to specify a 'dummy' level 1 with no random variation as in Chapter 4. If, however, we allow correlation between height and bone age then we will need to specify the model with no variation at level 1, the variances and covariance between bone age and height at level 2 and the between-individual variation at level 3.

Table 6.1 shows the fixed and random parameters for this model, omitting the estimates for the between-individual variation in the quadratic and cubic coefficients of the polynomial growth curve. We see that there are non-zero

Table 6.1 Height (cm) for adolescent growth, bone age, and adult height for a sample of boys. Age measured about 13.0 years. Level 2 variances and covariances shown; correlations in brackets

Parameter	Estimate (s.e.)
Fixed:	
Adult height intercept	174.4
Group (A–B)	0.25 (0.50)
Height intercept	153.0
Age	6.91 (0.20)
Age ^{2}	0.43 (0.09)
Age ^{3}	−0.14 (0.03)
Age ^{4}	−0.03 (0.01)
Age ^{5}	0.03 (0.03)
Bone age intercept	0.21 (0.09)
Age	0.03 (0.03)

Random:
Level 2

	Adult height	Height intercept	Age	Bone age intercept
Adult height	62.5			0.57 (0.08)
Height intercept	49.5 (0.85)	54.5		3.00 (0.44)
Age	1.11 (0.09)	1.14 (0.09)	2.5	0.02 (0.01)
Bone age intercept				0.85

Level 1 variance	
Height	0.89
Bone age	0.18

correlations between adult height and both the height growth and the bone age coefficients. This implies that the growth and bone age measurements can be used to make predictions of adult height. In fact, these predicted values are simply the estimated residuals for adult height. For a new individual, with information available at one or more ages on height or bone age, we simply estimate the adult height residual using the model parameters. Table 6.2 shows the estimated standard errors associated with predictions made on the basis of varying amounts of information. It is clear that the main gain in efficiency comes with the use of height with a smaller gain from the addition of bone age.

The method can be used for any measurements, either to be predicted or as predictors. In particular, covariates such as family size or social background can be included to improve the prediction. We can also predict other events of interest, such as the estimated age at maximum growth velocity.

Table 6.2 Standard errors for height predictions for specified combinations of height and bone age measurements

		Height measures (age)		
		None	11.0	11.0 12.0
Bone age measures				
None			4.3	4.2
11.0		7.9	3.9	3.8
11.0	12.0	7.9	3.7	3.7

Modelling an autocorrelation structure at level 1

6.4 So far we have assumed that the level 1 residuals are independent. In many situations, however, such an assumption would be false. For growth measurements the specification of level 2 variation serves to model a separate curve for each individual, but the between-individual variation will typically involve only a few parameters, as in the previous example. Thus, if measurements on an individual are obtained very close together in time, they will tend to have similar departures from that individual's underlying growth curve. That is, there will be 'autocorrelation' between the level 1 residuals. Examples arise from other areas, such as economics, where measurements on each unit, for example an enterprise or economic system, exhibit an autocorrelation structure and where the parameters of the separate time series vary across units at level 2.

A detailed discussion of multilevel time series models is given by Goldstein *et al* (1994). They discuss both the discrete time case, where the measurements are made at the same set of equal intervals for all level 2 units, and the continuous time case where the time intervals can vary. We shall develop the continuous time model here since it is both more general and flexible.

To simplify the presentation, we shall drop the level 1 and 2 subscripts and write a general model for the level 1 residuals as follows

$$\text{cov}(e_t e_{t-s}) = \sigma_e^2 f(s) \tag{6.6}$$

Thus, the covariance between two measurements depends on a variance function which depends on the time (or age) at which one of the measurements is made and the time difference between the measurements and another function of the time difference. The second function is conveniently described by a negative exponential reflecting the common assumption that with increasing time difference the covariance tends to a fixed value, $\alpha \sigma_e^2$, and typically this is assumed to be zero

$$f(s) = \alpha + \exp(-g(\beta, z, s)) \tag{6.7}$$

where β is a vector of parameters for explanatory variables z. Some choices for g are given in Table 6.3.

We can apply the methods described in Appendix 5.1 to obtain maximum likelihood estimates for these models, by writing the expansion

$$f(s,\beta,z) = \{1 + \sum_k \beta_{k,t} z_k g(H_t)\} f(H_t) - \sum_k \beta_{k,t+1} z_k g(H_t) f(H_t) \qquad (6.8)$$

so that the model for the random parameters is linear. Full details are given by Goldstein *et al* (1994).

A growth model with autocorrelated residuals

6.5   The data for this example consist of a sample of 26 boys each measured on nine occasions between the ages of 11 and 14 years (Harrison and Brush, 1990). The measurements were taken approximately 3 months apart. Table 6.4 shows the estimates from a model which assumes independent level 1 residuals with a constant variance. The model also includes a cosine term to model the seasonal variation in growth with time measured from the beginning of the year. If the seasonal component has amplitude α and phase γ we can write

$$\alpha \cos(t + \gamma) = \alpha_1 \cos(t) - \alpha_2 \sin(t)$$

In the present case the second coefficient is estimated to be very close to zero and is set to zero in the following model. This component results in an average growth difference between summer and winter estimated to be about 0.5 cm.

We now fit in Table 6.5 the model with $g = \beta_0 s$ which is the continuous time version of the first-order autoregressive model.

The fixed part and level 2 estimates are little changed. The autocorrelation parameter implies that the correlation between residuals three months (0.25 years) apart is 0.19.

Table 6.3 Some choices for the covariance function g for level 1 residuals

$g = \beta_0 s$	For equal intervals this is a first-order autoregressive series.
$g = \beta_0 s + \beta_1(t_1 + t_2) + \beta_2(t_1^2 + t_2^2)$	For time points t_1, t_2 this implies that the variance is a quadratic function of time.
$g = \begin{cases} \beta_0 s & \text{if no replicate} \\ \beta_1 & \text{if replicate} \end{cases}$	For replicated measurements this gives an estimate of measurement reliability $\exp(-\beta_1)$.
$g = (\beta_0 + \beta_1 z_{1j} + \beta_2 z_{2ij})s$	The covariance is allowed to depend on an individual level characteristic (e.g. gender) and a time-varying characteristic (e.g. season of the year or age).
$g = \begin{cases} \beta_0 s + \beta_1 s^{-1}, & s > 0 \\ 0, & s = 0 \end{cases}$	Allows a flexible functional form, where the time intervals are not close to zero.

Table 6.4 Height as a fourth degree polynomial on age, measured about 13.0 years. Standard errors in brackets; correlations in brackets for covariance terms

Parameter	Estimate (s.e.)		
Fixed:			
Intercept	148.9		
Age	6.19 (0.35)		
Age^2	2.17 (0.46)		
Age^3	0.39 (0.16)		
Age^4	−1.55 (0.44)		
Cos (time)	−0.24 (0.07)		
Random:			
Level 2			
	Intercept	Age	Age^2
Intercept	61.6 (17.1)		
Age	8.0 (0.61)	2.8 (0.7)	
Age^2	1.4 (0.22)	0.9 (0.67)	0.7 (0.2)
Level 1			
σ_e^2	0.20 (0.02)		

Table 6.5 Height as a fourth degree polynomial on age, measured about 13.0 years. Standard errors in brackets; correlations in brackets for covariance terms. Autocorrelation structure fitted for level 1 residuals

Parameter	Estimate (s.e.)		
Fixed:			
Intercept	148.9		
Age	6.19 (0.35)		
Age^2	2.16 (0.45)		
Age^3	0.39 (0.17)		
Age^4	−1.55 (0.43)		
Cos (time)	−0.24 (0.07)		
Random:			
Level 2			
	Intercept	Age	Age^2
Intercept	61.5 (17.1)		
Age	7.9 (0.61)	2.7 (0.7)	
Age^2	1.5 (0.25)	0.9 (0.68)	0.6 (0.2)
Level 1			
σ_e^2	0.23 (0.04)		
β	6.90 (2.07)		

Multivariate repeated measures models

6.6 We have already discussed the bivariate repeated measures model where the level 1 residuals for the two responses are independent. In the general multivariate case where correlations at level 1 are allowed, we can fit a full multivariate model by adding a further lowest level as described in Chapter 4. For the autocorrelation model this will involve extending the models to include cross-correlations. For example, for two response variables with the model of Table 6.5, we would write

$$g = \sigma_{e1}\sigma_{e2} \exp(-\beta_{12}s)$$

The special case of a repeated measures model where some or all occasions are fixed is of interest. We have already dealt with one example of this where adult height is treated separately from the other growth measurements. The same approach could be used with, for example, birthweight or length at birth. In some studies, all individuals may be measured at the same initial occasion and we can choose to treat this as a covariate rather than as a response. This might be appropriate where individuals were divided into groups for different treatments following initial measurements.

Scaling across time

6.7 For some kinds of data, for example educational achievement scores, different measurements may be taken over time on the same individuals so that some form of standardization may be needed before they can be modelled using the methods of this chapter. It is common in such cases to standardize the measurements so that at each measuring occasion they have the same population distribution. If this is done then we should not expect any trend in either the mean or variance over time, although there will still, in general, be between-individual variation. An alternative standardization procedure is to convert scores to age equivalents; that is, to assign to each score the age for which that score is the population mean or median. Where scores change smoothly with age this has the attraction of providing a readily interpretable scale. Plewis (1993) used a variant of this in which the coefficient of variation at each age is also fixed to a constant value. In general, different standardizations may be expected to lead to different inferences. The choice of standardization is, in effect, a choice about the appropriate scale along which measurements can be equated so that any interpretation needs to recognize this. A further discussion of this issue is given by Plewis (1994).

Cross-over designs

6.8 A common procedure for comparing the effects of two different treatments A, B, is to divide the sample of subjects randomly into two groups and then to assign A to one group followed by B, and B to the other group followed by A. The potential advantage of such a design is that the between-individual variation can be removed from the treatment comparison. A basic model for such a design with two treatments, repeated measurements on individuals and a single group effect can be written as follows

$$y_{ij} = \beta_0 + \beta_1 x_{1ij} + \beta_2 x_{2ij} + u_{0j} + u_{2j} x_{2ij} + e_{ij} \tag{6.9}$$

where X_1 is a dummy variable for time period and X_2 is a dummy variable for treatment. In this model we have not modelled the response as a function of time within treatment, but this can be added in the standard fashion described in previous sections. In the random part at level 2 we allow between-individual variation for the treatment difference and we can also structure the level 1 variance to include autocorrelation or different variances for each treatment or time period.

One of the problems with such designs is so-called 'carry over' effects whereby exposure to an initial treatment leaves some individuals more or less likely to respond positively to the second treatment. In other words, the u_{2j} may depend on the order in which the treatments were applied. To model this we can add an additional term to the random part of the model, say $u_{3j}\delta_{3ij}$, where δ_{3ij} is a dummy variable which is 1 when A precedes B and the second treatment is being applied, and zero otherwise. This will also have the effect of allowing level 2 variances to depend on the ordering of treatments. The extension to more than two treatment periods and more than two treatments is straightforward.

7

Multilevel Models for Discrete Response Data

Models for discrete response data

7.1   All the models of previous chapters have assumed that the response variable is continuously distributed. We now look at data where the response is essentially a count of events. This count may be the number of times an event occurs out of a fixed number of 'trials' in which case we usually deal with the resulting proportion as a response: an example is the proportion of deaths in a population, classified by age. We may have a vector of counts representing the numbers of events of different kinds which occur out of a total number of events: an example is given in Chapter 3 where we studied the number of responses to each, ordered, category of a question on abortion attitudes.

Statistical models for such data are referred to as 'generalized linear models' (McCullagh and Nelder, 1989). A 2-level model can be written in the general form

$$\pi_{ij} = f(X_{ij}\beta_j) \tag{7.1}$$

where π_{ij} is the expected value of the response for the ijth level 1 unit and f is a nonlinear function of the 'linear predictor' $X_{ij}\beta_j$. Note that we allow random coefficients at level 2. The model is completed by specifying a distribution for the *observed* response $y_{ij} | \pi_{ij}$. Where the response is a proportion, this is typically taken to be binomial and where the response is a count, it is taken to be Poisson. Equation (7.1) is a special case of the nonlinear model studied in Chapter 5 and we shall be using the results given there. It remains for us to specify the nonlinear 'link' function f. Table 7.1 lists some of the standard choices, with logarithms chosen to base e.

In addition to these we can also have the 'identity' function $f^{-1}(\pi) = \pi$, but this can create difficulties since it allows, in principle, predicted counts or proportions which are respectively less than zero or outside the range (0,1). Nevertheless, in many cases, using the identity function produces acceptable results which may differ little from those obtained with the nonlinear functions. In the following sections we consider each common type of model in turn with examples.

Table 7.1 Some nonlinear link functions

Response	$f^{-1}(\pi)$	Name
Proportion	$\log\{(\pi)/(1-\pi)\}$	logit
Proportion	$\log\{-\log(1-\pi)\}$	complementary log log
Vector of proportions	$\log(\pi_s/\pi_t)$ $(s=1,...,t-1)$	multivariate logit
Count	$\log(\pi)$	log

Proportions as responses

7.2   Consider the 2-level variance components model with a single explanatory variable, where the expected proportion is modelled using a logit link function

$$\pi_{ij} = \{1 + \exp(-[\beta_0 + \beta_1 x_{1ij} + u_{0j}])\}^{-1} \qquad (7.2)$$

The observed responses y_{ij} are proportions with the standard assumption that they are binomially distributed

$$y_{ij} \sim \mathrm{Bin}(\pi_{ij}, n_{ij}) \qquad (7.3)$$

where n_{ij} is the denominator for the proportion. We also have

$$\mathrm{var}(y_{ij} \mid \pi_{ij}) = \pi_{ij}(1-\pi_{ij})/n_{ij} \qquad (7.4)$$

We now write the model in the standard way including the level 1 variation as

$$y_{ij} = \pi_{ij} + e_{ij} z_{ij}, \quad z_{ij} = \sqrt{\pi_{ij}(1-\pi_{ij})/n_{ij}}, \quad \sigma_e^2 = 1 \qquad (7.5)$$

Using this explanatory variable Z, and constraining the level 1 variance associated with this to be one, we obtain the required binomial variance in Equation (7.4). When fitting a model we can also allow the level 1 variance to be estimated and, by comparing the estimated variance with the value 1.0, obtain a test for 'extra binomial' variation. Such variation may arise in a number of ways.

If we have omitted a level in the model, for example ignored household clustering in a survey with one or more individuals sampled from a household, we would expect a greater than binomial variation at the individual level. Likewise, suppose the individuals and households were nested within areas and we chose to classify individuals, say by gender and three social class groups, giving six cells in each area. If we treat these as the level 1 units so that the response is a proportion, then we no longer have a binomial variance since these proportions are based upon the sum of separate binomial variables with differing probabilities. In this case, the variance for cell j within an area would have the form

$$[E(\pi_j)(1 - E(\pi_j)) - \sigma_1^2]/n_j$$

where n_j is the cell size. To fit such a model we would specify an extra level 1 explanatory variable equal to $1/\sqrt{n_j}$ for the jth cell, with variance parameter at level 1, which was allowed to be negative (see Chapter 3). More generally, we can fit a model with an extra binomial parameter together with a further term, such as above, to give the following level 1 variance structure (omitting subscripts)

$$[\sigma_0^2 \pi (1 - \pi) + \sigma_1^2]/n$$

We do not, of course, know the true value of π_{ij} or π_j so that at each iteration we use estimates based upon the current values of the parameters. Because we are using only the mean and variance of the binomial distribution to carry out the estimation, the estimation is known as 'quasi-likelihood' (see Appendix 5.1).

Another way of modelling such extra binomial variation, which has certain advantages, is to insert a 'pseudo level' above level 1. Thus, for individuals sampled within households, level 1 would be that of the individual and we would specify level 2 as that of the individuals also to give exactly one level 1 unit per level 2 unit. We specify binomial variation at level 1, and at level 2 we can now fit further random coefficients. For example, if we fit a random coefficient for the explanatory variable with a variance that can be allowed to be negative, this is equivalent to specifying an extra level 1 variable $1/\sqrt{n_j}$ as above. In the above example, where individuals are classified by gender and social class we can create a level 2 unit coinciding with each level 1 unit, fit binomial variation at level 1 and add level 2 variation, which is a function of gender and social class, for example an additive function with four parameters (see Chapter 3). We may wish to model the between-area variation of the cell proportions in terms of a simple variance term, rather than as inversely proportional to n_j. In this case, we would choose a simple dummy variable structure rather than explanatory variables proportional to $1/\sqrt{n_j}$. This 'pseudo level' procedure is rather similar to the way in which a Meta Analysis with known level 1 variation is modelled (Chapter 3).

In Chapter 5, we made the distinction between models where the current level 2 residual estimates were added to the linear component of the nonlinear function when forming the Taylor expansion, in order to work with a linearized model, and those cases where they were not. The former is referred to as predictive quasi-likelihood (PQL) and the latter, marginal quasi-likelihood (MQL). In many applications, the MQL procedure will tend to underestimate the values of both the fixed and random parameters, especially where n_{ij} is small. In addition, we pointed out that greater accuracy is to be expected if the second-order approximation is used rather than the first-order based upon the first term in the Taylor expansion. Also, when the sample size is small the unbiased (RIGLS, REML) procedure should be used. Appendix 7.1 gives expressions for the second differentials required for the second-order procedure. To illustrate the difference, Table 7.2 presents the results of simulating the following model where the response is binary (0,1). The example assumes one moderate and one large level 2 variance.

$$\text{logit}(\pi_{ij}) = \beta_0 + \beta_1 x_{ij} + u_{0j}$$
$$y_{ij} \sim \text{Bin}(\pi_{ij}, 1)$$
$$\text{var}(u_{0j}) = 0.5, \ 1.0$$
$$\beta_0 = 0.5, \quad \beta_1 = 1.0$$

There are 50 level 2 units with 20 level 1 units in each level 2 unit. The following results are based upon 400 simulations of the above model for each variance value.

Table 7.2 Mean values of 400 simulations. Empirical standard error in first bracket; mean of estimated standard errors in second bracket (IGLS)

Parameter	True $\sigma_{u0}^2 = 0.5$		True $\sigma_{u0}^2 = 1.0$	
	MQL first order	PQL second order	MQL first order	PQL second order
σ_{u0}^2	0.386(0.115)(0.130)	0.480(0.157)(0.152)	0.672(0.157)(0.188)	0.964(0.278)(0.255)
β_0	0.448(0.126)(0.129)	0.499(0.139)(0.138)	0.420(0.145)(0.149)	0.500(0.171)(0.172)
β_1	0.934(0.154)(0.147)	1.018(0.168)(0.154)	0.875(0.147)(0.145)	1.017(0.171)(0.158)

Here, the denominator is 1.0 in all cases. It is clear that the MQL first-order model underestimates all the parameter values, whereas the second-order PQL model produces estimates close to the true values. The estimates given are based upon IGLS. In every case convergence was achieved in less than ten iterations. Very similar estimates for the fixed coefficients are obtained using RIGLS, and for the level 2 variances the PQL estimates become 0.498 and 0.996 respectively, which are even closer to the true values. In addition, the averages of the standard errors given by both models are reasonably close to those calculated empirically from the replications. If we calculate 95% confidence intervals for the parameters in the second-order PQL model using the estimated standard errors and assuming Normality then we find that, for the variance, about 91% of the intervals include the true value and that for β_0 and β_1, about 95% do so. Hence, inferences about the true values would not be too misleading. The results of Table 7.2 are based upon a balanced data set with equal numbers of level 1 units within each level 2 unit. Further, limited, simulations suggest that even where the data are very unbalanced, for example with some level 2 units containing only a single level 1 unit, the PQL second-order estimates remain close to the true values. These estimates appear to have good properties even with average observed probabilities as small as 0.1 or as large as 0.9 and a level 2 variance of 1.0 for the sample structure of this example.

More generally, when the average observed probability is very small (or very large), if many of the level 2 units have few level 1 units and there are very few level 2 units with large numbers of level 1 units, we will often find that where the response is binary, there will be many level 2 units where the responses are all zero. In such a case, convergence often may not be possible and, even where estimates are obtained, in general they will not be unbiased. This problem can be avoided by having a sufficient number of large level 2 units where there is adequate response heterogeneity, and in such cases we can obtain satisfactory estimates even where the average probabilities are very small or large. Further research into this issue is needed in order to make more precise recommendations. In all the following examples of this chapter we shall use the second-order PQL estimates, although in one case convergence could not be obtained so that second-order MQL estimates have been used.

An example from a survey of voting behaviour

7.3   The data were collected as part of a series of surveys carried out in Britain between 1964 and 1992 known as the British Election Studies (Heath *et al*, 1991). The respondents were interviewed following parliamentary general elections, and here we use the data from the elections which took place in 1983, 1987 and 1992. The response is whether the respondent voted for the Conservative party as opposed to the Liberal or Labour parties. The response is either one (voted Conservative) or zero, with the denominator always equal to one. Those who did not vote or voted for other parties are excluded. The level 2 unit is the year and the level 3 unit is the parliamentary constituency. Some constituencies were sampled in all three years. There are 8052 level 1 units, 780 level 2 units and 475 level 3 units. An alternative formulation is to specify a 2-level model, fitting variances and covariances for each year at level 2. This uses six parameters, however, as opposed to

Table 7.3 Weighted analysis of Conservative voting preference. Subscript v denotes level 3, subscript u denotes level 2. Analysis D is equally weighted. Binomial variation at level 1

Parameter	A	Model estimates (s.e.)		
		B	C	D
Intercept	0.173	0.188	0.153	0.172
Pet. Bourg. (Class 2)	0.50 (0.09)	0.49 (0.09)	0.63 (0.16)	0.51 (0.09)
Manual (Class 3)	−0.88 (0.05)	−0.91 (0.06)	−0.85 (0.09)	−0.89 (0.05)
1987	−0.05 (0.07)	−0.06 (0.07)	−0.04 (0.10)	−0.05 (0.07)
1992	0.02 (0.08)	0.01 (0.07)	0.11 (0.10)	−0.04 (0.08)
Interactions:				
Sc2 × 1987			−0.24 (0.21)	
Sc3 × 1987			0.02 (0.13)	
Sc2 × 1992			−0.14 (0.22)	
Sc3 × 1992			−0.21 (0.14)	
Random:				
σ_{v0}^2	0.37 (0.05)	0.37 (0.07)	0.36 (0.07)	0.38 (0.05)
σ_{v02}		−0.22 (0.09)	−0.21 (0.09)	
σ_{v2}^2		0.33 (0.20)	0.33 (0.21)	
σ_{v03}		−0.02 (0.06)	−0.02 (0.06)	
σ_{v23}		0.25 (0.11)	0.24 (0.11)	
σ_{v3}^2		0.18 (0.09)	0.19 (0.09)	
σ_u^2	0.04 (0.04)	0.03 (0.04)	0.03 (0.04)	0.05 (0.04)

two for the present model. In fact, the present model is equivalent to the 2-level model with the assumption of a constant covariance between years and equal between-constituency variance at each year. A preliminary test indicates that the 3-level model is an adequate fit.

The explanatory variables used are year and social class (classified as Non-manual, Petty Bourgeoisie and Manual, including foremen). Table 7.3 shows the results of three models of increasing complexity fitted to the data. Also, fitting extra-binomial variation at level 1 gives a variance estimate of 0.97 with a standard error of 0.16 indicating little departure from the binomial assumption.

In 1992 Scotland was oversampled so that each voter in Scotland had four times the probability of inclusion as one in the rest of Britain. Weighted analyses have been carried out with voters in 1983 and 1987 having a weight of 1.0, those in Scotland in 1992 having a weight of 0.28 and those in the rest of Britain in 1992 having a weight of 1.12 so that the average weight in 1992 is 1.0. For comparison, the last column in Table 7.3 shows the result of the unweighted (equally weighted) analysis. The 1992 estimate is now larger, reflecting the fact that, in 1992, Scotland was relatively more anti-Conservative. This illustrates the importance of weighting, as discussed in Chapter 3. If we include region as a factor in the model, with Wales as the base category, we see that the weighted and unweighted analyses again pro-

Table 7.4 Analysis of Conservative voting preference, including region; differentially weighted and equally weighted. Binomial variation at level 1

Parameter	Differentially weighted	Equally weighted
Intercept	−0.461	−0.483
Petit Bourg.	0.48 (0.09)	0.50 (0.09)
Manual	−0.85 (0.05)	−0.85 (0.05)
1987	−0.05 (0.07)	−0.05 (0.07)
1992	−0.005 (0.07)	0.02 (0.07)
Scotland	−0.05 (0.20)	0.07 (0.19)
North	0.33 (0.17)	0.34 (0.18)
Midlands	0.90 (0.18)	0.91 (0.18)
Southwest	1.00 (0.20)	1.01 (0.21)
Southeast	0.90 (0.17)	0.91 (0.18)
Random:		
σ^2_{v0}	0.24 (0.04)	0.24 (0.04)
σ^2_{u0}	0.025 (0.04)	0.043 (0.04)

duce different results. This time the unweighted analysis overestimates support for the Conservatives in 1992 overall and especially in Scotland, although, as in Table 7.3, the standard errors associated with these effects are large. Table 7.4 displays the results.

Figure 7.1 shows the Normal score plot for the constituency (level 3) residuals. The extreme values represent constituencies with very high or very low proportions of Conservative voters. The smaller slope of the line in Fig. 7.1 at the extremes indicates a smaller variation among these constituencies. Since such constituencies are typically associated with high proportions of non-manual and high proportions of manual voters respectively, we next fit a model where we allow different between-constituency variances for these groups, and the results are shown as analysis B of Table 7.3. With (0,1) binary data the likelihood ratio test statistic is unreliable and so we carry out an approximate test on the random parameters for the null hypothesis that the additional variation for social class groups 2 and 3 is zero. We obtain an approximate chi squared of 13.3 on five degrees of freedom corresponding to a P value of 0.02. The between-constituency variance is the same for social class groups 1 and 2 (0.36) but greater for the manual group (0.54). The remaining parameters in the model are little changed and a normal residual plot for the basic constituency residuals shows a somewhat more linear relationship.

For the fixed coefficients, in analysis B a test for equality of year effects produces a non-significant result ($\chi^2_2 = 0.77$). It is possible, however, that there is an interaction between social class and year; that is, there are year differences *within* social class groups. Analysis C shows the result of fitting the appropriate interaction terms. A test for the significance of these gives a chi squared of 4.9 with four degrees of freedom, so that there does, in fact, seem to be little evidence of any interaction.

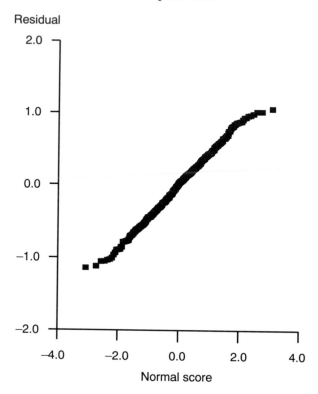

Figure 7.1   Normal residual plot for constituencies: Analysis A in Table 7.3.

Models for multiple response categories

7.4   In this section we extend the model for a single proportion as the outcome to the case of a set of proportions, for example the proportions voting for all three political parties in the example of the previous section. The response is now multivariate and we can define a generalization of the ordinary logit model to define a multivariate logit as follows for a simple 2-level variance components model

$$\log\left(\frac{\pi_{ij}^{(s)}}{\pi_{ij}^{(t)}}\right) = \beta_0^{(s)} + \beta_1^{(s)}x_{ij} + u_j^{(s)}, \quad s = 1,\ldots,t-1 \tag{7.6}$$

where there are t response categories. Choosing one category (t) as the base category avoids redundancy and a singular covariance matrix and hence the need to introduce generalized inverses into the estimation. There are cases where this procedure is inappropriate and we discuss these below. Thus, (7.6) specifies the model for each of the remaining $t-1$ categories with $\sum_{h=1}^{t}\pi_{ij}^{(h)} = 1$. When $t = 2$ this reduces to the ordinary logit model.

We treat the $t-1$ response categories as a multivariate response vector as described in Chapter 4 using dummy variables with no variation at level 1 and the true level 1 covariance matrix specified at level 2. For example, in the case of the three response categories of the voting behaviour example, $t = 3$ and we specify a bivariate model where level 2 describes the between-individual variation. If we make the standard assumption that the observed response proportions follow a multinomial distribution then the level 2 covariance matrix has the form

$$
n_{ij}^{-1}
\begin{pmatrix}
\pi_{ij}^{(1)}(1-\pi_{ij}^{(1)}) & & & \\
-\pi_{ij}^{(1)}\pi_{ij}^{(2)} & \ddots & & \\
\vdots & & \ddots & \\
-\pi_{ij}^{(1)}\pi_{ij}^{(t-1)} & \cdots & \cdots & \pi_{ij}^{(t-1)}(1-\pi_{ij}^{(t-1)})
\end{pmatrix}
\tag{7.7}
$$

where n_{ij} is the total number of responses over all categories. In the voting behaviour example this is always one since each individual votes for just one party.

We can create the covariance structure (7.7) as follows. Define the explanatory variables

$$
\left.
\begin{aligned}
z_{1ij} &= \sqrt{\pi_{ij}/n_{ij}}, & z_{2ij} &= \pi_{ij}/\sqrt{2n_{ij}} \\
z_{3ij} &= -\pi_{ij}/\sqrt{2n_{ij}}, & \pi_{ij} &= \{\pi_{ij}^{(s)}\}
\end{aligned}
\right\}
\tag{7.8}
$$

and specify Z_1 to have a random coefficient at level 1 with variance constrained to 1.0 and Z_2, Z_3 to have random coefficients at level 2 constraining their variances to zero and their covariance to 1.0. This produces the structure (7.7) and extra multinomial variation can be achieved by allowing the variance and covariance to be different from 1.0 but constraining them to be equal. Level 3 then defines variation between higher level units, for example years or constituencies.

The response vector itself is not restricted to a single classification. Thus, suppose we had a response which was an individual's first voting preference cross-classified by their second preference. This produces nine response categories of which just one contains the value 1 for each individual. A 'main effects' model extension to (7.6) would express the probability of any particular combination of first and second preferences as an additive function of a term for the first and a term for the second preference, as follows

$$
\log\left(\frac{\pi_{ij}^{(s=s_1,s_2)}}{\pi_{ij}^{(t)}}\right) = \beta_0^{(s_1)} + \beta_0^{(s_2)} + \beta_1^{(s_1)}x_{1ij} + \beta_1^{(s_2)}x_{2ij} + u_j^{(s_1)} + u_j^{(s_2)}, \quad s = 1,\ldots,t-1
$$

For the random parameters it would be reasonable to attempt to fit a model where the covariances between the $u_j^{(s_1)}$, $u_j^{(s_2)}$ were zero in order to reduce the number of random parameters in the model.

To see how we can interpret the parameters of these models we write

$$
\log(\pi_{ij}^{(r)}/\pi_{ij}^{(s)}) = (\beta_0^{(r)} - \beta_0^{(s)}) + (\beta_1^{(r)} - \beta_1^{(s)})x_{ij} + (u_j^{(r)} - u_j^{(s)})
\tag{7.9}
$$

so that a unit change in x_{ij} multiplies the ratio of the rth and sth response probabilities by $\exp(\beta_r^{(r)} - \beta_1^{(s)})$. Likewise, a difference of d in the residuals or in the intercept terms multiplies this ratio by e^d.

This formulation of the multicategory response model is adequate for models such as (7.6) where coefficients are fitted for each response category (except the base). There are other models, however, where we may wish to fit a function defined *across the categories*. This will often be the case when there are a large number of ordered categories where we wish to study linear, quadratic, etc trends across the categories; although, as we point out later, there will often be more satisfactory procedures for such cases based upon consideration of the *cumulative* probabilities $\pi_{ij}^{(1)}$, $\pi_{ij}^{(1)} + \pi_{ij}^{(2)}$,....

Where we do wish to treat the categories symmetrically and define a function across the response categories, we replace the intercept term $\beta_0^{(s)}$ in (7.6) by such a function. If we assume a linear function then (7.6) can be written as

$$
\log\left(\frac{\pi_{ij}^{(s)}}{\pi_{ij}^{(t)}}\right) = \gamma_0 + \gamma_1 w^{(s)} + (\beta_0 + \beta_1 w^{(s)})x_{ij} + u_j^{(s)}, \quad s = 1,...,t-1 \qquad (7.10)
$$

where $w^{(s)}$ is the score assigned to category s. We might also wish to structure the level 2 variation, for example writing $u_j^{(s)} = u_{0j} + u_{1j}w^{(s)}$. Such a model will be especially useful where the number of categories becomes large.

In (7.10) the choice of base category is no longer irrelevant since the score assigned to this category does not appear in the model. We can avoid this difficulty by defining the multivariate logistic over all the response categories ($s = 1,...,t$) and in (7.10) the level 2 resulting covariance matrix will not be singular so long as the set of response category probabilities is predicted using fewer responses than there are categories. An alternative formulation, using the Poisson with a log link function as described below, will often be more convenient.

An example of voting behaviour with multiple responses

7.5   We now look at the voting data, with the response being the 3-category choice of party: Conservative, Liberal or Labour. Table 7.5 gives the results of the analysis using the same explanatory variables as in Table 7.3 but omitting year at level 2. We have chosen Labour, 1983 and Non Manual as the base categories. The estimation uses the second-order approximation without residuals. It was not possible to obtain convergence for the procedure using the expansion about the current residual estimates.

The results for the Conservative voters are broadly in line with those from the analysis in Table 7.3. For the Liberals we see that there is greater support among the manual social class than the petit bourgeoisie and relatively more support in 1987 than either 1983 or 1992. At the constituency level there appears to be little correlation between the Conservative and Liberal support. It should be remembered, however, that we have only fitted a variance-components model at constituency level and there may be more substantial correlations within social class groups or in different years, but we shall not pursue this further.

Models for counts

7.6   Instead of using a set of proportions as the response we can consider the underlying event counts as the set of responses. Thus, for example, in the voting data,

Table 7.5 Analysis of Conservative and Liberal voting preferences

	Model estimates (s.e.)	
	Conservative	Liberal
Intercept	0.86 (0.06)	0.22 (0.06)
Pet. Bourg.	0.29 (0.08)	−0.36 (0.16)
Manual	−1.07 (0.05)	0.46 (0.09)
1987	−0.09 (0.06)	0.20 (0.10)
1992	−0.25 (0.07)	−0.13 (0.10)
Random:		
Level 2		
Variance	0.27 (0.04)	0.24 (0.04)
Covariance	0.007 (0.03)	

suppose we classify individuals by three social-class and three year categories. In each of the nine cells within each level 2 unit we have counts of the numbers voting for each party, which yields 27 counts. The expected number of individuals voting for each party can be written

$$m_{sij} = M_j \pi_{ij}^{(s)}$$

where s indexes the three parties, i indexes the nine cells within each level 2 unit and M_j is the number of individuals in the jth level 2 unit. Our inferences are therefore conditional on these totals. We write, corresponding to (7.6)

$$\log(m_{sij}) = \log(M_j) + \beta_0^{(s)} + \beta_1^{(s)} x_{ij} + u_j^{(s)}, \quad s = 1, \ldots t \qquad (7.11)$$

The term $\log(M_j)$ is a fixed part offset and when using such offsets it may be better to centre them about their mean in order to avoid numerical instabilities. Corresponding to the multinomial assumption, we now make the assumption of a Poisson distribution for the *observed* counts n_{sij}, which are assumed conditionally independent with

$$E(n_{sij}) = m_{sij}, \quad \text{var}(n_{sij} \mid m_{sij}) = m_{sij}$$

For the voting data we can now define a two-level model where, at level 2, we have the constituency, and the level 1 units are the set of counts for the classification of party by year and social class. A basic additive model will have explanatory variables consisting of an intercept, two dummy variables for party, two dummy variables for year and two for social class. We would normally also wish to include interactions between party and year and party and social class.

The level 1 variation is specified using the predicted number for each level 1 unit and the estimation follows the same pattern as for the binomial model, using the corresponding expressions given in Table 5.1. The level 1 random part will be defined by a dummy variable equal to the square root of the predicted count and with variance constrained to one where a Poisson distribution is assumed.

There are some applications where the response is a count and we do not require an offset, or where the offset is effectively constant. For example, if we were interested in the number of times individuals visited their general practitioner or physician in a year, we could collect data over a one-year period for all individuals and study the variation in counts across practitioners (level 2) according to individual and practitioner characteristics.

There are variations on the Poisson distribution assumption which we may wish to use. For example, the negative binomial distribution can be obtained from a process whereby the response is generated by counting the number of incidents for each level 1 unit and where, conditional on the fitted explanatory variables and higher level terms, the mean count for each level 1 unit has a gamma distribution with index v. This leads us to consider level 1 variance functions of the general form $k_1 m + k_2 m^2$, where $k_1 = 1$ gives the negative binomial distribution with $k_2 = 1/v$. We could add further terms or consider even a nonlinear function.

Ordered responses

7.7   In Chapter 3 we analysed a study where the response was a scale where the score ranged in value from 0 to 7, that is, there were eight ordered categories. Such response scales are common and, as in our example, are often analysed by assigning scores and then treating them as if they were continuous. While this may often be satisfactory, there are situations, for example where the distribution is very skew, where such a procedure is questionable. One possible alternative, mentioned in the preceding section, is to assign scores to the categories of the response variable and then carry out an analysis based upon the multinomial or Poisson model, using all the response categories in the analysis. Such a procedure, however, typically relies on choosing a suitable scoring system, just as does the continuous response model. One possibility in these cases is to assign scores by minimizing a measure of between-unit disagreement as in correspondence analysis or dual scaling (Greenacre, 1984; Goldstein, 1987c). In this section we shall look at procedures which avoid any of the arbitrariness of assumptions involved when assigning numerical scores.

To exploit the ordering we shall base our models upon the *cumulative* response probabilities rather than the response probabilities for each category. We define these as

$$E(y_{ij}^{(s)}) = \gamma_{ij}^{(s)} = \sum_{h=1}^{s} \pi_{ij}^{(h)}, \quad s = 1,\ldots,t-1 \tag{7.12}$$

where $y_{ij}^{(s)}$ are the observed proportions out of a total n_{ij} and s now indexes the ordered cumulative categories. If we assume an underlying multinomial distribution for the category probabilities the cumulative proportions have a covariance matrix given by

$$\operatorname{cov}(y_{ij}^{(s)}, y_{ij}^{(r)}) = \gamma_{ij}^{(s)}(1 - \gamma_{ij}^{(r)})/n_{ij}, \quad s \le r \tag{7.13}$$

We can therefore fit models to these cumulative proportions (or counts conditional on a fixed total) in the same way as with the multinomial response vector, substituting the covariance matrix (7.13) for (7.7). A discussion of these and related models is given in McCullagh and Nelder (1989).

A common model choice is the *proportional odds* model which uses a logit link namely

$$\gamma_{ij}^{(s)} = \{1 + \exp - [\alpha^{(s)} - (X\beta)_{ij}]\}^{-1} \qquad (7.14)$$

where the negative coefficient of $(X\beta)_{ij}$ implies that increasing values of this linear component are associated with increasing probabilities with increasing s. We also require $\alpha^{(1)} \leq \alpha^{(2)} \ldots \leq \alpha^{(t-1)}$.

Another choice is the *proportional hazards* model which uses a log log link to give

$$\gamma_{ij}^{(s)} = \{1 - \exp - [\alpha^{(s)} - (X\beta)_{ij}]\} \qquad (7.15)$$

An important special case of these models is where the categories are ordered in time so that $\alpha^{(s)}$ can be modelled as a function of time, and satisfy the above order relationship among these parameters. Some choices would be

$$\alpha^{(s)} = \delta \log(t_s), \quad \alpha^{(s)} = \delta t_s \qquad (7.16)$$

Such a model might be used in developmental studies where individuals pass through a set of time-ordered stages. In studies of children, for example, it is possible to identify 'milestones' of development through which children pass, starting with none until all have been passed when developmental 'maturity' is reached. A repeated measures study would count the number passed at each time point so yielding a cumulative proportion in relation to time and other covariates. We would then be able to fit a 2-level model with variation between individuals involving any of the parameters in (7.14), (7.15) or (7.16). In the extreme case, with just a single milestone, these models are equivalent to the event duration models we consider in Chapter 9.

Another example of longitudinal discrete response data is where, at each measurement occasion, we have a vector of ordered categorical responses and each individual in the study responds to one category. The cumulative response vector for each individual at each occasion then contains zero, for each response category less than the category to which the individual responds, and a one for that category and each higher one. We can model the time dependence within the set of explanatory variables X, and we would normally wish to include the possibility of interactions between the $\alpha^{(s)}$ and time. In such a model the basic covariance structure given by (7.13) represents the between-occasion covariation. Thus, although the data structure is represented by level 1 as the categories, level 2 as occasion and level 3 as individual, the higher level variation is only estimated at level 3. This can be compared with the simple binary response model where the binomial response variance is that between occasions, and the structure defines occasion as level 1 and individual as level 2 since there is a single response for each occasion. We also note that similar considerations apply to all the multicategory response models, with higher level variation estimated at level 3 and above, as pointed out in Section 7.4.

Mixed discrete–continuous response models

7.8 An extension of the multivariate models considered in Chapter 4 is where some of the responses are continuous and some are discrete. For example, in a repeated measures study we may have a response which is the (discrete) maturity stage that an individual has reached, as well as continuous measurements such as height and weight. In some circumstances we may wish to treat, say, the maturity stage as the response, conditional on height and weight and further covariates, including age. In other situations, for example if we are interested in prediction systems as in Chapter 6, then we would wish to consider all the measurements as responses, conditional on covariates. In another example, suppose we have measurements on smoking habit, including whether someone smoked and if so at what rate. We can consider this as a bivariate response model where each individual has a binary response for whether or not they smoke, and if they do a further response for the number smoked per day.

We shall develop the model for the case of individual smoking habits with one binary and one continuous response and then look at the more general case of several binary responses. The extension to several responses of each type is straightforward as is the extension to multicategory responses and count data.

As in the standard multivariate multilevel model we have no variation at level 1 and at level 2, that of the individual, we have a binomial variance associated with the smoking/no smoking response and a between-individual variance for the number smoked. The variance for the binary response is the usual binomial variance and that for the continuous response is a parameter to be estimated. At higher levels, the variances and covariances will be defined in the standard fashion using the linearization procedure for the binomial response. For a 2-level model with individuals nested within, say, households we write the following model

$$y_{ijk} = \delta_i \exp\{1 + \exp[(-X_1\beta_1)_{ijk} + u_{1k}]\}^{-1}$$
$$+ \delta_i e_{1jk} + (1 - \delta_i)[(X_2\beta_2) + u_{2k} + e_{2jk}] \qquad (7.17)$$
$$\delta_i = 1 \text{ if binary, } 0 \text{ if continuous}$$

If the binary response for a smoker is one, and the continuous response for a non-smoker is zero, the covariance for the jkth individual between the binary and continuous response is given by $(1 - \hat{\pi}_{jk})\hat{\pi}_{jk}\hat{y}_{jk}$. In the more general case an individual can have any combination of responses, as in the maturity example, and the individual level covariance will have the form of an (adjusted) biserial covariance $(1 - \hat{\pi}_{jk})\hat{\pi}_{jk}(\hat{y}_{1jk} - \hat{y}_{2jk})$, where $\hat{\pi}_{jk}$ is the estimated probability of a positive response, and $\hat{y}_{1jk}, \hat{y}_{2jk}$ are, respectively, the predicted values of the continuous response for a positive and negative binary response. From the model we will obtain an estimate of this covariance and $\hat{y}_{jk} = \hat{\pi}_{jk}\hat{y}_{1jk} + (1 - \hat{\pi}_{jk})\hat{y}_{2jk}$ from which the separate predictions can be calculated.

A particular case of interest is where we have two binary or proportion responses. Suppose, for example, that in an educational survey we know the proportion of students in each school who pass an English exam and also the proportion who pass a Mathematics exam, but we have no information about how many pass or fail one or both. In other words, for the 2 × 2 table containing the numbers in each pass/fail category we only have the numbers in the margins. The level 2 covariance, in terms of the predicted proportions, has the form $\hat{\pi}_{(11)jk} - \hat{\pi}_{(1)jk}\hat{\pi}_{(2)jk}$ so that, with estimates of the marginal probabilities available from the model, the level 2 covariance estimate allows us to obtain an estimate of the joint probability of success on both Mathematics and English for a given set of explanatory variables.

Note that the procedure depends upon the assumption of binomial variation. Of course, if we had all the original information then we would fit a model where there was a response for each cell of the table.

This approach may also be of use where separate surveys are conducted within the same level 2 units and each one produces a proportion as a response. If there is overlap between the samples used, then there will exist level 2 covariances, and if information about the detailed nature of the overlap is available it will be possible, in principle, to obtain estimates of the joint probabilities.

APPENDIX 7.1

Differentials for some discrete response models

The Logit–Binomial model

$$f = [1 + \exp(-X\beta)]^{-1}$$
$$f' = f[1 + \exp(X\beta)]^{-1}$$
$$f'' = f'[1 - \exp(X\beta)][1 + \exp(X\beta)]^{-1}$$

The Logit–Multinomial (Multivariate Logit) model

$$f^{(s)} = \exp(X\beta^{(s)})[1 + \sum_{h=1}^{t-1} \exp(X\beta^{(h)})]^{-1}, \quad s = 1,\ldots,t-1$$

$$f'^{(s)} = f^{(s)}[1 + \sum_{h \neq s} \exp(X\beta^{(h)})][1 + \sum_{h=1}^{t-1} \exp(X\beta^{(h)})]^{-1}$$

$$f''^{(s)} = f'^{(s)} \exp(X\beta^{(s)})[1 + \sum_{h=1}^{t-1} \exp(X\beta^{(h)})]^{-1}$$

The Log–Poisson model

$$f = \exp(X\beta)$$
$$f' = \exp(X\beta)$$
$$f'' = \exp(X\beta)$$

The log log–Binomial model

$$f = 1 - \exp[-\exp(X\beta)]$$
$$f' = (1 - f)\exp(X\beta)$$
$$f'' = f'[1 - \exp(X\beta)]$$

8

Multilevel Cross Classification

Random cross classifications

8.1 In previous chapters we have considered only data where the units have a purely hierarchical or nested structure. In many cases, however, a unit may be classified along more than one dimension. An example is students classified both by the school they attend and by the neighbourhood where they live. We can represent this diagrammatically, as in Fig. 8.1, for three schools and four neighbourhoods with between one and six students per school/neighbourhood cell. The cross classification is at level 2 with students at level 1.

	School 1	School 2	School 3
Neighbourhood 1	x x x x	x x	x
Neighbourhood 2	x	x x x x x x	x x x
Neighbourhood 3	x x	x	x x x x
Neighbourhood 4	x x x	x x	x x

Figure 8.1 A random cross classification at level 2.

Another example is in a repeated measures study where children are measured by different raters at different occasions. If each child has its own set of raters not shared with other children then the cross classification is at level 1, occasions by raters, nested within children at level 2. This can be represented diagrammatically, as in Fig. 8.2, for three children with up to seven measurement occasions and up to three raters per child. We see that the cross classification takes place entirely within the level 2 units; child 2 having only one rater, while for child 3 each rater rates only once. We note that, by definition, a level 1 cross classification has only one unit per cell.

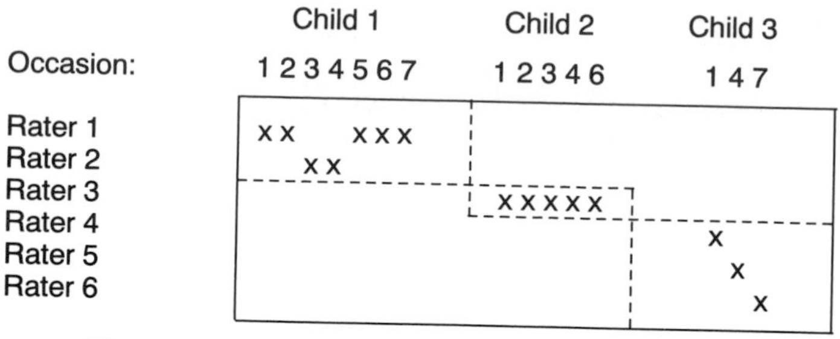

	Child 1	Child 2	Child 3
Occasion:	1 2 3 4 5 6 7	1 2 3 4 6	1 4 7

Figure 8.2   A random cross classification at level 1.

If now the same set of raters is involved with all the children the crossing is at level 2, as can be seen in Fig. 8.3, with three raters and three children and up to five occasions.

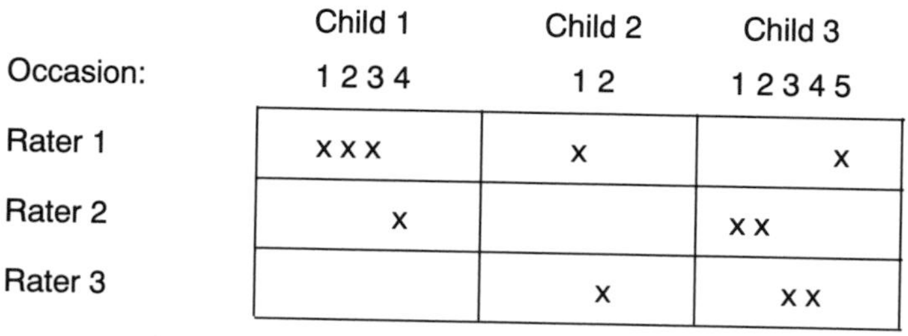

	Child 1	Child 2	Child 3
Occasion:	1 2 3 4	1 2	1 2 3 4 5
Rater 1	x x x	x	x
Rater 2	x		x x
Rater 3		x	x x

Figure 8.3   A random cross classification at level 2.

Figure 8.3 is formally the same structure as Fig. 8.1 with the level 1 variance being that between occasions.

These basic cross classifications occur commonly when a simple hierarchical structure breaks down in practice. Consider, for example, a repeated measures design which follows a sample of students over time, say once a year, within a set of classes for a single school. We assume first that each class group is taken by the same teacher. The hierarchical structure is then a three level one with occasions grouped within students who are grouped within classes. If we had several schools then schools would constitute the level 4 units. Suppose, however, that students change classes during the course of the study. For three students, three classes and up to three occasions we might have the pattern in Fig. 8.4.

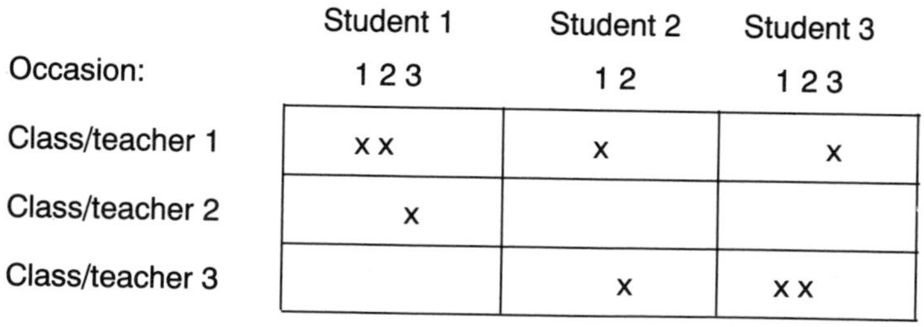

	Student 1	Student 2	Student 3
Occasion:	1 2 3	1 2	1 2 3
Class/teacher 1	x x	x	x
Class/teacher 2	x		
Class/teacher 3		x	x x

Figure 8.4   Students changing classes/teachers.

Formally, this is the same structure as Fig. 8.3; that is, a cross classification at level 2 for classes by students. Such designs will also occur in panel or longitudinal studies of individuals who move from one locality to another, or workers who change their place of employment. If we now include schools, these will be classified as level 3 units, but if students also change schools during the course of the study then we obtain a level 3 cross classification of students by schools with classes nested at level 2 within schools and occasions as the level 1 units. The students have moved from being crossed with classes to being crossed with schools. Note that since students are crossed at level 3 with schools they are also automatically crossed with any units nested within schools and we do not need separately to specify the crossing of classes with students.

Suppose now that, instead of the same teachers taking the classes throughout the study, the classes are taken by a completely new set of teachers every year and new groupings of students are formed each year too. Such a structure with four different teachers at two occasions for three students is given in Fig. 8.5.

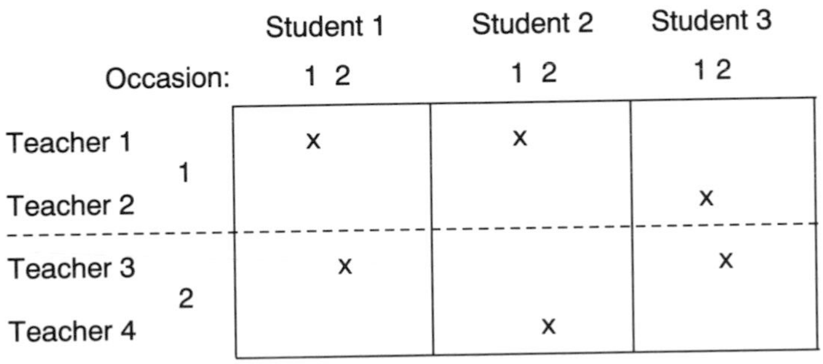

Figure 8.5   Students changing teachers and groups.

This is now a cross classification of teachers by students at level 2 with occasion as the level 1 unit. We note that most of the cells are empty and that there is at most one level 1 unit per cell so that no independent between-occasion variance can be estimated as pointed out above. In fact we can also view this as a level 1 cross classification of teachers by students, with missing data, and occasion can be modelled in the fixed part, for example using a polynomial function of age. Raudenbush (1993) gave an example of such a design, and provided details of an EM estimation procedure for 2-level 2-way cross classifications with worked examples.

We can have a design which is a mixture of those given by Fig. 8.4 and Fig. 8.5 where some teachers are retained and some are new at each occasion. In this case we would have a cross classification of teachers by students at level 2 where some of the teachers only had observations at one occasion. More generally, we can have an unbalanced design where each teacher is present at a variable number of occasions. Other examples of such designs occur in panel studies of households where, over time, some households split up and form new households. The total set of all households is crossed with individuals at level 2 with occasion at level 1. The households which remain intact for more than one occasion provide the information for estimating level 1 variation.

With two occasions where we have the same teachers or intact groups we can

Occasion 2

	Teacher 1	Teacher 2	Teacher 3
Teacher 1	x x x x x	x	x x
Occasion 1 Teacher 2	x x	x x x x	
Teacher 3	x	x x x	x x x x

Figure 8.6   Teachers cross classified by themselves at two occasions.

formulate an alternative cross classification design, which may be more appropriate in some cases. Instead of cross classifying students by teachers we consider cross classifying the set of all teachers at the first occasion by the same set at the second occasion, as in Fig. 8.6.

We have 22 students who are nested within the cross classification of teachers at each occasion. The difference between this design and that in Fig. 8.4 is analogous to the difference between a two-occasion longitudinal design where a second occasion measurement is regressed on a first occasion measurement and the two-occasion repeated measures design where a measurement is related to age or time. In Fig. 8.6 we are concerned with the contribution from each occasion to the variation in, say, a measurement made at occasion 2. In Fig. 8.4 on the other hand, although we could fit a separate between-teacher variance for each occasion, the response variable is essentially the same one measured at each occasion. Designs such as that of Fig. 8.6 are useful where, for example, measurements are made on the same set of students and schools at the start and end of schooling, as in school effectiveness studies, and where students can move between schools. In such cases we may also wish to introduce a 'weight' to reflect the time spent in each school, and we shall discuss this below.

We now set out the structure of these basic models and then go on to consider extensions and special cases of interest.

A basic cross classified model

8.2   Goldstein (1987a) sets out the general structure of a model with both hierarchical and cross classified structures. We consider first the simple model of Fig. 8.1 with variance components at level 2 and a single variance term at level 1.

We shall refer to the two classifications at level 2 using the subscripts j_1, j_2 and, in general, parentheses will group classifications at the same level. We write the model as

$$y_{i(j_1 j_2)} = X_{i(j_1 j_2)}\beta + u_{j_1} + u_{j_2} + e_{i(j_1 j_2)} \tag{8.1}$$

The covariance structure at level 2 can be written in the following form

$$\left. \begin{array}{l} \text{cov}(y_{i(j_1 j_2)} y_{i'(j_1 j_2')}) = \sigma_{u_1}^2 \\[2mm] \text{cov}(y_{i(j_1 j_2)} y_{i'(j_1' j_2)}) = \sigma_{u_2}^2 \\[2mm] \text{var}(y_{i(j_1 j_2)}) = \text{cov}(y_{i(j_1 j_2)} y_{i'(j_1 j_2)}) = \sigma_{u_1}^2 + \sigma_{u_2}^2 \end{array} \right\} \tag{8.2}$$

Thus, the level 2 variance is the sum of the separate classification variances, the covariance for two level 1 units in the same classification is equal to the variance for that classification and the covariance for two level 1 units which do not share either classification is zero. If we have a model where random coefficients are included for either or both classifications, then analogous structures are obtained. We can also add further ways of classification with obvious extensions to the covariance structure.

Appendix 8.1 shows how cross classified models can be specified and estimated efficiently using a purely hierarchical formulation, and we can summarize the procedure using the simple model of Fig. 8.1. We specify one of the classifications, most efficiently the one with the larger number of units, as a standard hierarchical level 2 classification. For the other classification we define a dummy (0,1) variable for each unit which is one if the observation belongs to that unit and zero if not. Then we specify that each of these dummy variables has a coefficient random at level 3 and, in addition, constrain the resulting set of level 3 variances to be equal. The variance estimate obtained is that required for this classification and the level 2 variance for the other classification is the one we require for that.

If we have a third classification at level 2 then we can obtain the third variance by defining a similar set of dummy variables with coefficients varying at level 4 and variances constrained to be equal. This procedure generalizes straightforwardly to sets of several random coefficients for each classification, with dummy variables defined as the products of the basic (0,1) dummy variables used in the variance components case and with corresponding variances and covariances constrained to be equal within classifications. In general, a p-way cross classification at any level can be modelled by inserting sets of random variables at the next $p-1$ higher levels. Thus, in a 2-level model with two crossed classifications at level 1 we would obtain a three-level model with the original level 2 at level 3 and the level 1 cross classifications occupying levels 1 and 2.

Examination results for a cross classification of schools

8.3 The data consist of scores on school leaving examinations obtained by 3435 students who attended 19 secondary schools cross classified by 148 primary schools in Fife, Scotland (Paterson, 1991). Before their transfer to secondary school at the age of 12 each student obtained a score on a verbal reasoning test, measured about the population mean of 100 and with a population standard deviation of 15.

The model is as follows

$$y_{i(j_1j_2)} = \beta_0 + \beta_1 x_{1i(j_1j_2)} + u_{j_1} + u_{j_2} + e_{i(j_1j_2)} \tag{8.3}$$

and the results are given in Table 8.1. Random coefficients for verbal reasoning were also fitted but the coefficients are estimated as zero.

Ignoring the verbal reasoning score we see that the between-primary school variance is estimated to be more than three times that between secondary schools. The principal reason for this is that the secondary schools are, on average, far larger than primary schools, so that within a secondary school, secondary school differences are averaged. Such an effect will often be observed where one classification has far fewer units than another, for example where a small number

Table 8.1 Analysis of examination scores by secondary and primary school attended. The subscript 1 refers to primary and 2 to secondary school

Parameter	Estimate (s.e.) A	Estimate (s.e.) B	Estimate (s.e.) C
Fixed:			
Intercept	5.50	5.98	5.99
Verbal reasoning	–	0.16 (0.003)	0.16 (0.003)
Random:			
$\sigma^2_{u(1)0}$	1.12 (0.20)	0.27 (0.06)	–
$\sigma^2_{u(2)0}$	0.35 (0.16)	0.011 (0.021)	0.28 (0.06)
σ^2_e	8.1 (0.2)	4.25 (0.10)	4.26 (0.10)

of schools is crossed with a large number of small neighbourhoods, or a small number of teachers is crossed with a large number of students at level 1 within schools. In such circumstances we need to be careful about our interpretation of the relative sizes of the variances.

When the verbal reasoning score is added to the fixed part of the model the between secondary school variance becomes very small; the between primary school variance is also considerably reduced as is the level 1 variance. The third analysis shows the effect of removing the cross classification by primary school. The between secondary school variance is now only a little smaller than in analysis A without the verbal reasoning score. Using analysis C alone, which is typically the case with school effectiveness studies which control for initial achievement, we would conclude that there were important differences between the progress made in secondary schools. From analysis B, however, we see that most of this is explained by the primary schools attended. Of course, the verbal reasoning score is only one measure of initial achievement, but these results illustrate that adjusting for achievement at a single prior time may not be adequate.

Computational considerations

8.4   Analysis A in Table 8.1 took about 40 seconds per iteration on a 66 MHz 486 PC using the ML3 software, approximately ten times longer than analysis C. This relative slowness is due to the size of the single level 3 unit which contains all the 3435 level 1 units. For very much larger problems the computing considerations will become of greater concern, so that some procedure for speeding up the computations would be useful.

In the present analysis there are 120 cells of the cross classification which contain only one student. If we eliminate these from the analysis we obtain two disjoint subsets containing 14 and 5 secondary schools. There are a further 24 cells containing two students and if these are removed we obtain six disjoint subsets the largest of which contains eight secondary schools. Table 8.2 shows the estimates from the resulting analyses.

Table 8.2 Examination scores for secondary and primary school classification omitting small cells

Parameter	Estimate (s.e.) $\leqslant$ 1 student	Estimate (s.e.) $\leqslant$ 2 students
Fixed:		
Intercept	6.00	6.00
Verbal reasoning	0.16 (0.003)	0.16 (0.003)
Random:		
$\sigma_{u(1)0}^2$	0.27 (0.06)	0.25 (0.06)
$\sigma_{u(2)0}^2$	0.004 (0.021)	0.028 (0.030)
σ_e^2	4.28 (0.11)	4.29 (0.11)

The only substantial difference is in the between secondary school variance which is anyway poorly estimated. The first analysis took about 15 seconds and the second about six seconds. Such computational advantages in some cases may well outweigh a slight loss in precision.

Interactions in cross classifications

8.5   Consider the following extension of (8.1)

$$y_{i(j_1 j_2)} = X_{i(j_1 j_2)}\beta + u_{j_1} + u_{j_2} + u_{(j_1 j_2)} + e_{i(j_1 j_2)} \tag{8.4}$$

We have now added an 'interaction' term to the model which was previously an additive one for the two variances. The usual specification for such a random interaction term is that it has simple variance $\sigma_{u(12)}^2$ across all the level 2 cells (Searle *et al*, 1992). To fit such a model we would define each cell of the cross classification as a level 2 unit with a between cell variance $\sigma_{u(12)}^2$, a single level 3 unit with a variance σ_{u1}^2 and a single level 4 unit with a variance σ_{u2}^2. The adequacy of such a model can be tested against an additive model using a likelihood ratio test criterion. For the example in Table 8.1 this interaction term is estimated as zero. While this indicates that the cross classification is adequate, because the between secondary school variance is so small we would not expect to be able to detect such an interaction.

Extensions to this model are possible by adding random coefficients for the interaction component, just as random coefficients can be added to the additive components. For example, the gender difference between students may vary across both primary and secondary schools in the example of Section 8.3 and we can fit an extra variance and covariance term for this to both the additive effects and the interaction.

Level 1 cross classifications

8.6   Some interesting models occur when units are basically cross classified at level 1. By definition we have a design with only one unit per cell, as shown for example in Fig. 8.2, and we can also have a level 2 cross classification which is formally equivalent to a level 1 cross classification where there is just one unit per cell as in Fig. 8.5. This case should be distinguished from the case where a level 2 cross classification happens to produce no more than one level 1 unit in a cell as a result of sampling, so that the confounding occurs by chance rather than by design.

If the level 1 additive variation model is inadequate we can fit an interaction term as in (8.4) and this will give the model (with simple variation at level 2)

$$y_{(i_1 i_2)j} = X_{(i_1 i_2)j}\beta + u_j + e_{i_1 j} + e_{i_2 j} + e_{(i_1 i_2)j} \tag{8.5}$$

where for level 1 we use a straightforward extension of the notation for a level 2 cross classification. To specify this model we would define the u_j as random at level 4, the $e_{i_1 j}, e_{i_2 j}$ as random at levels 3 and 2, each with a single unit and the interaction term random across the cells of the cross classification at level 1, within the original level 2 units.

Suppose now that we were able to extend the design by replicating measurements for each cell of the level 1 cross classification. Then (8.5) would refer to a 3-level model with replications as level 1 units, and which could be written as follows, where the subscript h denotes replications

$$y_{h(i_1 i_2)j} = X_{h(i_1 i_2)j}\beta + u_j + e_{i_1 j} + e_{i_2 j} + e_{h(i_1 i_2)j} \tag{8.6}$$

Since (8.5) is just model (8.6) with one unit per cell, we could interpret the interaction variance in (8.5) as an estimate of the within-cell variance where confounding has occurred by chance and where the assumption of an additive model is reasonable.

So-called 'generalizability theory' models (Cronbach and Webb, 1975) can be formulated as level 1 cross classifications. The basic model is one where a test or other instrument consisting of a set of items, for example ratings or questions, is administered to a sample of individuals. The individuals are therefore cross classified by the items at level 1 and may be further nested within schools etc at higher levels. In educational test settings the item responses are often binary so that we would apply the methods of Chapter 7 to the present procedures in a straightforward way.

Cross-unit membership models

8.7   In some circumstances units can be members of more than one higher level unit at the same time. An example is friendship patterns where, at any time, individuals can be members of more than one friendship group. Another example is where children belong to more than one 'extended' family, which includes aunts and uncles as well as parents. In an educational system students may attend more than one institution. In all such cases we shall assume that for each higher level unit to which a lower level unit belongs there is a known weight, summing to 1.0 for each lower level unit, which represents, for example, the amount of time spent in that unit.

We may also have data where, although there is no cross-unit membership, there is some uncertainty about which higher level unit some lower level units belong to. For example, in a survey of students, information about their neighbourhood of residence may only be available for a few students for larger geographical units. For these cases it may be possible to assign a weight for each of the constituent neighbourhoods which is, in effect, a probability of belonging to each, based upon available information. Such a structure can be analysed formally as a cross-unit membership model with most students having a single weight of 1.0 and the remainder zero.

Consider the 2-level variance components model (8.1) with each level 1 unit belonging to at most two level 2 units where the j_1, j_2 subscripts now refer to the same type of unit.

$$\left.\begin{aligned} y_{i(j_1 j_2)} &= X_{i(j_1 j_2)}\beta + w_{1ij_1}u_{j_1} + w_{2ij_2}u_{j_2} + e_{i(j_1 j_2)} \\ w_{1ij_1} &+ w_{2ij_2} = 1 \end{aligned}\right\} \tag{8.7}$$

The overall contribution at level 2 is therefore the weighted sum over the level 2 units to which each level 1 unit belongs. This leads to the following covariance structure

$$\text{var}(y_{i(j_1 j_2)}) = (w_{1ij_1}^2 + w_{2ij_2}^2)\sigma_u^2 + \sigma_e^2$$

$$\text{cov}(y_{i(j_1 j_2)}y_{i'(j_1 j_2)}) = (w_{1ij_1}w_{1i'j_1} + w_{2ij_2}w_{2i'j_2})\sigma_u^2$$

$$\text{cov}(y_{i(j_1 j_2)}y_{i'(j_1' j_2)}) = w_{2ij_2}w_{2i'j_2}\sigma_u^2$$

This has the structure of a standard 2-level cross classified model with the additional constraint $\sigma_{u1}^2 = \sigma_{u2}^2 = \sigma_u^2$ and where the explanatory indicator variables Z_1, Z_2 described in Fig. 8.1.1 in Appendix 8.1 have the value 1 replaced by the relevant weights for each level 1 unit. As with the standard cross classification this model can be extended to include random coefficients and general p-unit membership (see Appendix 8.1).

Multivariate cross classified models

8.8   For multivariate models the responses may have different structures. Thus, in a bivariate model, one response may have a 2-level hierarchical structure and the other may have a cross classification at level 2. Suppose, for example that we measure the height and the mathematics attainment of a sample of students from a sample of schools. The mathematics attainment is assessed by a different set of teachers in each school and the heights are measured by a single anthropometrist. For the mathematics scores there is a level 1 cross classification of students within each school whereas for height there is a 2-level hierarchy with students nested within schools. Height and mathematics attainment will be correlated at both the student and the school level and we can write a model for this structure as follows

$$\left.\begin{aligned} y_{h(i k_2)j} &= \delta_{1h}(X_{1(i k_2)j}\beta_1 + u_{1j} + e_{1ihj} + e_{1k_2j}) + \delta_{2h}(X_{2ihj}\beta_2 + u_{2j} + e_{2ihj}) \\ \text{cov}(u_{1j}u_{2j}) &= \sigma_{u12} \quad \text{cov}(e_{1ihj}e_{2ihj}) = \sigma_{e12} \\ \delta_{1h} &= 1 \text{ if mathematics, 0 if height,} \quad \delta_{2h} = 1 - \delta_{1h} \end{aligned}\right\} \tag{8.8}$$

where all other covariances are zero. This will therefore be specified as a 4-level model with the bivariate structure as level 1 and level 2 units being individual students. There will be a single level 3 unit with the coefficients of the dummy variables for teachers having variances random at this level, with level 4 being that of the school.

Finally, we have already mentioned that cross classified models can have a discrete response and the models of Chapter 7 can be fitted. We can also fit, for example, time series models as discussed in Chapter 6 and, in general, cross classified structures can incorporate all the types of models that can be fitted for purely hierarchical structures.

APPENDIX 8.1

Random cross classified data structures

We illustrate the procedure using a 2-level model with a crossing at level 2.

The 2-level cross classified model, using the notation in Appendix 2.1, can be written

$$y_{i(j_1 j_2)} = X_{i(j_1 j_2)}\beta + \sum_{h=1}^{q_1} z_{1hij_1} u_{1hj_1} + \sum_{h=1}^{q_2} z_{2hij_2} u_{2hj_2} + e_{i(j_1 j_2)} \qquad (8.1.1)$$

Parentheses group the ways of classification at each level. We have two sets of explanatory variables, type 1 and type 2, for the random components defined by the columns of $Z_1(n \times p_1 q_1)$, $Z_2(n \times p_2 q_2)$ where p_1, p_2 are, respectively, the number of categories of each classification.

$$Z_1 = \{z_{1hij_1}\}, \quad Z_2 = \{z_{2hij_2}\}$$

$z_{1hij_1} = z_{1him}$ if $j_1 = m$, for mth type 1 level 2 unit, 0 otherwise

$z_{2hij_2} = z_{2him}$ if $j_2 = m$, for mth type 2 level 2 unit, 0 otherwise

These variables are dummy variables where for each level 2 unit of type 1 we have q_1 random coefficients with covariance matrix $\Omega_{(1)2}$ and likewise for the type 2 units. To simplify the exposition we restrict ourselves to the variance component case where we have

$$\left. \begin{array}{l} \Omega_{(1)2} = \sigma^2_{(1)2}, \quad \Omega_{(2)2} = \sigma^2_{(2)2} \\ E(\tilde{Y}\tilde{Y}^T) = V_1 + Z_1(\sigma^2_{(1)2} I_{(p_1)})Z_1^T + Z_2(\sigma^2_{(2)2} I_{(p_2)})Z_2^T \end{array} \right\} \qquad (8.1.2)$$

Consider Fig. 8.1 in Chapter 8 where schools are ordered within neighbourhoods. The explanatory variables will have the structure shown in Fig. 8.1.1 for the first eight students. It is clear that the second term in (8.1.2) can be written as

$$Z_1(\sigma^2_{(1)2} I_{(p_1)})Z_1^T = J\sigma^2_{(1)2} J^T$$

where J is an $(n \times 1)$ vector of ones. The third term is of the general form $Z_3\Omega_3 Z_3^T$, namely a level 3 contribution where, in this case, there is only a single level 3 unit,

i,j_1	i,j_2	Z_{11}	Z_{12}	Z_{13}	Z_{14}	Z_{21}	Z_{22}	Z_{23}
1,1	1,1	1	0	0	0	1	0	0
2,1	2,1	1	0	0	0	1	0	0
3,1	3,1	1	0	0	0	1	0	0
4,1	4,1	1	0	0	0	1	0	0
5,1	1,2	1	0	0	0	0	1	0
6,1	2,2	1	0	0	0	0	1	0
7,1	1,3	1	0	0	0	0	0	1
1,2	2,1	0	1	0	0	1	0	0

Figure 8.1.1   Explanatory variables for level 2 cross classification of Fig. 8.1.

with no covariances between the random coefficients of the Z_{2h} and with the variance terms constrained to be equal to a single value, $\sigma^2_{(2)2}$.

More generally, we can specify a level 2 cross classified variance components model by modelling one of the classifications as a standard hierarchical component and the second as a set of dummy explanatory variables, one for each category, with the random coefficients uncorrelated and with variances constrained to be equal. If this second (type 2) classification has further explanatory variables with random coefficients as in (8.1.1) then we form extended dummy variable 'interactions' as the product of the basic dummy variables and the further explanatory variables with random coefficients, so that these coefficients have variances and covariances within the same type 2 level 2 unit but not across units. In addition, the corresponding variances and covariances are constrained to be equal.

To extend this to further ways of classification we add levels. Thus, for a three way cross classification at level 2 we choose one classification, typically that with the largest number of categories, to model in standard hierarchical fashion at level 2; the second to model with coefficients random at level 3 as above; and the third to model in a similar fashion with coefficients random at level 4. The same principle applies to cross classifications at level 1 nested within level 2 units. The level 1 cross classification is modelled as a 2-level hierarchy with the original level 2 units becoming level 3 units. We can also allow simultaneous crossing at more than one level. Thus, for example, if there is a 2-way cross classification at level 1 and a 3-way cross classification at level 2, we will require five levels, the first two describing the level 1 cross classification and the next three describing the level 2 cross classification.

Chapter 8 discusses the level 2 cross unit membership model where level 1 units can belong to more than one level 2 unit with predetermined weights. Because the structure imposed above level 2 replicates that at level 2, we need only in fact specify a single level 2 unit with explanatory variable design matrix Z containing dummy weight vectors and Ω_u as diagonal of order equal to the number of level 2 units, and elements equal to σ^2_u.

9

Multilevel Event History Models

Event history models

9.1   This class of models, also known as survival time models or event duration models, has as the response variable the length of time between 'events'. Such events may be, for example, birth and death, or the beginning and end of a period of employment with corresponding times being length of life or duration of employment. There is a considerable theoretical and applied literature, especially in the field of biostatistics, and a useful summary is given by Clayton (1988). We consider two basic approaches to the modelling of duration data. The first is based upon 'proportional hazard' models. The second is based upon direct modelling of the log duration, often known as 'accelerated life models'. In both cases we may wish to include explanatory variables.

The multilevel structure of such models arises in two general ways. The first is where we have repeated durations within individuals, analogous to our repeated measures models of Chapter 5. Thus, individuals may have repeated spells of various kinds of employment of which unemployment is one. In this case we have a 2-level model with individuals at level 2, often referred to as a renewal process. We can include explanatory dummy variables to distinguish these different kinds of employment or states. The second kind of model is where we have a single duration for each individual, but the individuals are grouped into level 2 units. In the case of employment duration the level 2 units would be firms or employers. If we had repeated measures on individuals within firms then this would give rise to a 3-level structure.

Censoring

9.2   A characteristic of duration data is that for some observations we may not know the exact duration but only that it occurred within a certain interval (known as interval censored data), was less than a known value (left censored data), or greater than a known value (right censored data). For example, if we know at the time of a study, that someone entered their present employment before a certain date then the information available is only that the duration is longer than a known value. Such data are known as right censored. In another case we may know that someone entered and then left employment between two measurement occasions, in which case we know only that the duration lies in a known interval. The models described in this chapter have procedures for dealing with censoring. In the case of the parametric models, where there are relatively large proportions of censored data, the assumed form of the distribution of duration lengths is important, whereas in the partially parametric models the distributional form is ignored. It is assumed that the censoring mechanism is non-informative; that is, independent of the duration lengths.

In some cases, we may have data that are censored but where we have no duration information at all. For example, if we are studying the duration of first marriage and we end the study when individuals reach the age of 30, all those marrying for the first time after this age will be excluded. To avoid bias we must therefore ensure that age of marriage is an explanatory variable in the model and report results conditional on age of marriage.

There is a variety of models for duration times. In this chapter we show how some of the more frequently used models can be extended to handle multilevel data structures. We consider first hazard-based models.

Hazard-based models in continuous time

9.3　The underlying notions are those of *survivor* and *hazard* functions. Consider the (single level) case where we have measures of length of employment on workers in a firm. We define the proportion of the workforce employed for periods greater than t as the *survivor function* and denote it by

$$S(t) = 1 - F(t) = 1 - \int_0^t f(u)du$$

where $f(t)$ is the density function of length of employment. The *hazard* function is defined as

$$h(t) = f(t)/S(t)$$

and represents the instantaneous risk, in effect the (conditional) probability of someone who is employed at time t, ending employment in the next (small) unit interval of time.

The simplest model is one which specifies an exponential distribution for the duration time, $f(t) = \lambda e^{-\lambda t}$ ($t \geqslant 0$) which gives $h(t) = \lambda$, so that the hazard rate is constant and $S(t) = e^{-\lambda t}$. In general, however, the hazard rate will change over time and a number of alternative forms have been studied (see, for example, Cox and Oakes, 1984). A common one is based on the assumption of a Weibull distribution, namely

$$g(t) = (\alpha/t)e^{\alpha \ln(t)+\delta} \; \exp(-e^{\alpha \ln(t)+\delta})$$

or the associated extreme value distribution formed by replacing t by $u = e^t$.

Another approach to incorporating time-varying hazards is to divide the time scale into a number of discrete intervals within which the hazard rate is assumed constant; that is, we assume a piecewise exponential distribution. This may be useful where there are 'natural' units of time, for example based on menstrual cycles in the analysis of fertility, and this can be extended by classifying units by other factors where time varies over categories. We will discuss such discrete time models in a later section.

The most widely used models, to which we shall devote our discussion, are those known as *proportional* hazards models, and the most common definition is $h(t;\eta) = \lambda(t)e^{\eta}$. The term η denotes a linear function of explanatory variables which we shall model explicitly in Section 9.5. It is assumed that $\lambda(t)$, the baseline hazard function, depends only on time and that all other variation between units is incorporated into the linear predictor η. The components of η may also depend upon time, and in the multilevel case some of the coefficients will also be random variables.

Parametric proportional hazard models

9.4 For the case where we have known duration times and right censored data, define the cumulative baseline hazard function $\Lambda(t) = \int_0^t \lambda(u)du$ and a variable w with mean $\mu = \Lambda(t)e^{\eta}$, taking the value one for uncensored and zero for censored data. It can be shown (McCullagh and Nelder, 1987) that the maximum likelihood estimates required are those obtained from a maximum likelihood analysis for this model where w is treated as a Poisson variable. This computational device leads to the loglinear Poisson model for the ith observation

$$\ln(\mu_i) = \ln(\Lambda(t_i)) + \eta_i \tag{9.1}$$

where the term $\Lambda(t_i)$ is treated as an offset, that is, a known function of the linear predictor.

The simplest case is the exponential distribution, for which we have $\Lambda(t) = \lambda t$. Equation (9.1) therefore has an offset $\ln(t_i)$ and the term $\ln(\lambda)$ is incorporated into η. We can model the response Poisson count using the procedures of Chapter 6, with coefficients in the linear predictor chosen to be random at levels 2 or above. This approach can be used with other distributions. For the Weibull distribution, of which the exponential is a special case, the proportional hazards model is equivalent to the log duration model with an extreme value distribution and we shall discuss its estimation in a later section.

The semiparametric Cox model

9.5 The most commonly used proportional hazard models are known as *semiparametric proportional hazard models* and we now look at the multilevel version of the most common of these in more detail.

Consider the 2-level proportional hazard model for the jkth level 1 unit

$$h(t_{jk}; X_{jk}) = \lambda(t_{jk})\exp(X_{jk}\beta_k) \tag{9.2}$$

where X_{jk} is the row vector of explanatory variables for the level 1 unit and some or all of the β_k are random at level 2. We adopt the subscripts j, k for levels one and two for reasons which will be apparent below.

We suppose that the times at which a level 1 unit comes to the end of its duration period or 'fails' are ordered and at each of these we consider the total 'risk set'. At failure time t_{jk} the risk set consists of all the level 1 units which have been censored or for which a failure has not occurred immediately prior to t_{jk}. Then the ratio of the hazard for the unit which experiences a failure and the sum of the hazards of the remaining risk set units is

$$\frac{\exp(X_{j'k'}\beta_{k'})}{\sum_{j,k} \exp(X_{jk}\beta_k)}$$

which is simply the probability that the failed unit is the one denoted by j', k' (Cox, 1972). It is assumed that, conditional on the X_{jk}, these probabilities are independent.

Several procedures are available for estimating the parameters of this model (see for example Clayton, 1991, 1992). For our purposes it is convenient to adopt the following, which involves fitting a Poisson or equivalent multinomial model of the kind discussed in Chapter 7.

At each failure time l we define a response variate for each member of the risk set

$$y_{ijk(l)} = \begin{cases} 1 \text{ if } i \text{ is the observed failure} \\ 0 \text{ if not} \end{cases}$$

where i indexes the members of the risk set, and j, k level 1 and level 2 units. If we think of the basic 2-level model as one of employees within firms then we now have a 3-level model where each level 2 unit is a particular employee and contains n_{jk} level 1 units where n_{jk} is the number of risk sets to which the employee belongs. Level 3 is the firm. The explanatory variables can be defined at any level. In particular, they can vary across failure times, allowing so-called time-varying covariates. Overall proportionality, conditional on the random effects, can be obtained by ordering the failure times across the whole sample. In this case the *marginal* relationship between the hazard and the covariates generally is not proportional. Alternatively, we can consider the failure times ordered only *within* firms, so that the model yields proportional hazards within firms. In this case, we can structure the data as consisting of firms at level 3, failure times at level 2 and employees within risk sets at level 1. In both cases, because we make the assumption of independence across failure times within firms, the Poisson variation is at level 1 and there is no variation at level 2. In other words we can collapse the model to two levels, within firms and between firms.

A simple variance components model for the expected Poisson count is written as

$$\pi_{jk(l)} = \exp(\alpha_l + X_{jk}\beta + u_k) \tag{9.3}$$

where there is a 'blocking factor' α_l for each failure time. In fact, we do not need generally to fit all these nuisance parameters: instead we can obtain efficient estimates of the model parameters by modelling α_l as a smooth function of the time points, using, say, a low-order polynomial or a spline function (Efron, 1988) .

For the model which assumes overall proportionality an estimator of the baseline surviving fraction for an individual in the kth firm at time h, where $X_{jk} = 0$, is

$$\hat{S}_h = \exp(-\sum_{l \leqslant h} e^{\hat{\alpha}_l + \hat{u}_k})$$

and the estimate for an individual with specific covariate values X_{jk} is

$$\hat{S}_h^{\{\exp(X_{jk}\beta\,)\}} \tag{9.4}$$

For the model which assumes proportionality within firms these two expressions become respectively

$$\hat{S}_h = \exp(-\sum_{l \leqslant h} e^{\hat{\alpha}_l}), \quad \hat{S}_h^{\{\exp(X_{jk}\beta)+\hat{u}_k\}}$$

Note that where there are ties, the offset described below must be added to the estimates $\hat{\alpha}_l$ or $\hat{\alpha}_l + \hat{u}_k$ in the above expressions. Where we fit polynomials to the blocking factors, the $\hat{\alpha}_l$ are estimated from the polynomial coefficients, and the surviving fraction can be plotted against the time associated with each interval.

Tied observations

9.6 We have assumed so far that each failure time is associated with a single failure. In practice many failures will often occur at the same time, within the accuracy of measurement. Sometimes, data may also be deliberately grouped in time. Such failures are referred to as 'ties' and are conveniently dealt with by fitting a Poisson model with an offset which is just $\ln(n_j)$ or $\ln(n_{jk})$, where n_j, n_{jk} are the number of tied observations associated with the jth or jkth unit, depending on whether a marginal relationship is being fitted or not. Of course, in the case of distinct failure times this offset is zero. This procedure for handling ties is equivalent to that described by Peto (1972) (see also McCullagh and Nelder, 1989).

Repeated measures proportional hazard models

9.7 As in the case of ordinary repeated measures models described in Chapter 6 we can consider the case of multiple episodes or durations within individuals with between and within individual variation and possibly further levels where individuals may be nested within firms, etc. The models of previous sections can be applied to such data, but there are further considerations which arise. Where each individual has the same fixed number n of episodes, we can treat these, as in Chapter 5, as constituting n variates so that we have an n-variate model with an $(n \times n)$ covariance matrix between individuals. The variates may be either really distinct measurements or simply the different episodes in a fixed ordering. This is the model considered by Wei *et al* (1989) who defined proportionality as *within* individuals. We can also model a multivariate structure where, within individuals, there are repeated episodes for a number of different types of interval. For each type of interval we may have coefficients random at the individual level and these coefficients will generally also covary at that level.

Often with repeated measures models, the first episode is different in nature from subsequent ones. An example might be the first episode of a disease which may tend to be longer or shorter than subsequent episodes. If the first episode is treated as if it were a separate variate then the subsequent episodes can be regarded as having the same distribution, as in the previous section.

Another possible complication in repeated measures data, as in Chapter 5, is that we may not be able to assume independence between durations within individuals. This will then lead to serial correlation models which can be estimated using the procedures discussed in Chapter 6 for the parametric log duration models discussed below.

Example using birth interval data

9.8   The data are a series of repeated birth intervals for 379 Hutterite women living in North America (Larsen and Vaupel, 1993; Egger, 1992). The response is the length of time in months from birth to conception, ranging from 1 to 160, with the first birth interval ignored and no censored information. This gives 2235 births in all.

There is information available on the mother's birth year, her age in years at the start of the birth interval, whether the previous child was alive or dead, and the duration of marriage at the start of the birth interval. Since we have a large number of women each with a relatively small number of intervals we have assumed overall proportionality, with failure times ordered across the whole sample. Table 9.1 gives the results for a variance components analysis and one where several random coefficients are estimated. A fourth-order polynomial was adequate to smooth the blocking factors.

Table 9.1 Proportional hazards model for Hutterite birth intervals. In the random part subscript 0 refers to intercept, 1 to previous death

Parameter	Estimate (s.e.) A	Estimate (s.e.) B
Fixed:		
Intercept	−3.65	−3.64
Mother's birth year − 1900	0.026 (0.003)	0.026 (0.003)
Mother's age (year − 20)	−0.008 (0.014)	−0.004 (0.014)
Previous death	0.520 (0.118)	0.645 (0.144)
Marriage duration (months)	−0.003 (0.001)	−0.004 (0.001)
Random:		
σ^2_{u0}	0.188 (0.028)	0.188 (0.028)
σ_{u01}		0.005 (0.088)
σ^2_{u1}		0.381 (0.236)

The only coefficient estimated with a non-zero variance at level 2 was whether or not the previous birth died, and a large sample chi-squared test for the two random parameters for this coefficient gives a *P*-value of 0.01 on 2 degrees of freedom. An increase on the linear scale is associated with a shorter interval. Thus, the birth interval decreases for the later born mothers and also if the previous birth is a death. The interval is somewhat shorter the longer the marriage duration with little additional effect of maternal age. This apparent lack of a substantial age effect seems to be a consequence of the high correlation (0.93) between duration of marriage and age. Higher order terms for duration and age were fitted but the estimated coefficients were small and not significant at the 10% level. The between-individual standard deviation is about 0.4 which is comparable in size with the effect of a previous death. The between-individual standard deviation for a model which fits no covariates is 0.45 so that the covariates explain only a small proportion of the between-individual variation. Figure 9.1 shows two average estimated surviving fraction curves for a woman aged 20, born in 1900 with marriage duration 12 months. The higher curve is for where there was a previous live birth and the lower where there was a previous death.

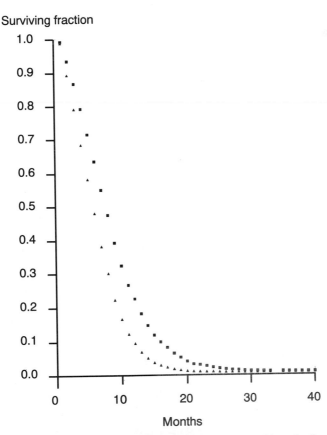

Figure 9.1 Probability of exceeding each birth interval length; live birth ■, previous death ▲.

The discrete time (piecewise) proportional hazards model

9.9   Where time is grouped into preassigned categories we can write the survivor function at time interval l, the probability that failure occurs after this interval, as s_l. This gives

$$f_l = s_{l-1} - s_l, \quad h_l = f_l / s_{l-1}, \quad s_0 = 1$$

This gives

$$s_l = \prod_{t=1}^{l} (1 - h_t)$$

which can be used to estimate the survivor function from a set of estimated hazards.

For the proportional hazards model (9.2) and a 2-level model the expected hazard is given (Aitkin *et al*, 1989) by

$$\pi_{jk(l)} = 1 - \exp(-e^{X_{jk}\beta_k + \alpha_{(l)}})$$

$$\log\{-\log(1 - \pi_{jk(l)})\} = X_{jk}\beta_k + \alpha_{(l)}$$

where, as before, the $\alpha_{(l)}$ are constants to be estimated, one for each time interval. This leads to a model where the response is a binomial variate, being the number of deaths divided by the number in the risk set at the start of the interval (see also Egger, 1992). Any censored observations in an interval are excluded from the risk set. The estimation follows that for the logit binomial model described in Chapter 7, except that we now require the first and second differentials of the log log function, namely

$$f' = \exp(\pi - e^{\pi}), \quad f'' = (1 - e^{\pi})\exp(\pi - e^{\pi})$$

As in the Cox model, we can fit a polynomial function to the successive time intervals, rather than the full set of blocking factors. The data will be ordered within level 2 units so that a risk set in general will extend over several such units. A general procedure is to specify the response for each level 1 unit as binary, that is zero if the unit survives the interval and one if not, with the appropriate $\alpha_{(l)}$ in the fixed part. Thus, a 2-level model will become specified as a 3-level model with the binomial variation at level 1 and the actual level 1 units at level 2. The model can be further extended to polytomous outcomes, or 'competing risks', where several different kinds of failure can occur. The analysis follows the same pattern, but with the response being a multinomial variate, and the corresponding models of Chapter 7 can be applied with a different linear predictor for each outcome category.

Log duration models

9.10   For the accelerated life model the distribution function for duration is commonly assumed to be of the form

$$f(t; X, \beta) = f_0(te^{X\beta})e^{X\beta}$$

where f_0 is a baseline function (Cox and Oakes, 1984).

For a 2-level model this can be written as

$$l_{ij} = \ln(t_{ij}) = X_{ij}\beta_j + e_{ij} \tag{9.5}$$

which is in the standard form for a 2-level model. We shall assume Normality for the random coefficients at level 2 (and higher levels) but at level 1 we shall study other distributional forms for the e_{ij}. The level 1 distributional form is important where there are censored observations. We first consider the common choice of an extreme value distribution for the log duration L, conditional on $X_{ij}\beta_j$ which, as we noted above, implies an equivalence with the proportional hazards model. Omitting level subscripts we write

$$f(l; \alpha, \delta) = \alpha\, e^{-\alpha l + \delta} \exp(-e^{-\alpha l + \delta}) \quad -\infty < l < \infty$$

$$E(L) = \alpha^{-1}(\delta - \gamma), \quad \mathrm{var}(L) = \frac{\pi^2}{6\alpha^2}, \quad \gamma = 0.5772 \tag{9.6}$$

For (9.5) this gives

$$\pi_{ij} = \mathrm{Pr}(L > l_{ij}) = 1 - \exp(-e^{-\alpha l_{ij} + \delta_{ij}})$$

$$\pi'_{ij} = \alpha \cdot \exp\{-e^{\alpha l_{ij} + \delta_{ij}}\} e^{-\alpha l_{ij} + \delta_{ij}} \tag{9.7}$$

The mean of L is incorporated into the fixed predictor. If we have no censored data we estimate the parameters for the model given by (9.5) by treating it as a standard multilevel model. We note that the estimation is strictly quasi-likelihood since we are using only the mean and variance properties of the level 1 distribution. If we assume a simple level 1 variance then we can iteratively estimate α from the above relationship and we also obtain for the 2-level model (9.5)

$$\delta_{ij} = \gamma + \alpha(X_{ij}\beta_j)$$

Where there is complex variation at level 1, then α will vary with the level 1 units. To estimate the survival function for a given level 2 unit we first condition on the covariates and random coefficients, that is $X_{ij}\beta_j$, and then use (9.7).

We can choose other distributional forms for the log duration distribution. These include the log gamma distribution, the Normal and the logistic. Thus, for example, for the Normal distribution we have

$$\pi_{ij} = 1 - \Phi(z_{ij})$$

$$\pi'_{ij} = [\sigma_e \sqrt{(2\pi)}]^{-1} \phi(z_{ij})$$

$$z_{ij} = [l_{ij} - (X\beta)_{ij}] / \sigma_e$$

where Φ, ϕ are the cumulative and density functions of the standard Normal distribution. Quasi-likelihood estimates can be obtained for any suitable distribution with two parameters. The possibility of fitting complex variation at level 1 can be expected to provide sufficient flexibility using these distributions for most purposes.

Censored data

9.11   Where data are censored in log duration models, we require the corresponding probabilities. Thus, for right censored data we would use (9.7) with the corresponding formulae for interval or left censored data. For each censored observation we therefore have an associated probability, say π_{ij}, with the response variable value of one.

This leads to a bivariate model, in which for each level 1 unit either the response is the continuous log duration time or it takes the value one if censored with corresponding explanatory variables in each case. There are basically two explanatory variables for the level 1 variation, one for the continuous log duration response and one for the binomial response. In the former case we can extend this for complex level 1 variation, as in the example analysis below. For the latter we use the standard logit model as described in Chapter 7, possibly allowing for extra-binomial variation. The random parameters at level 1 for the two components are uncorrelated. When carrying out the computations, we may obtain starting values for the parameters using just the uncensored observations.

Since the same linear function of the explanatory variables enters into both the linear and nonlinear parts of this model, we require only a single set of fixed part explanatory variables, although these will require the appropriate transformation for the logit response as described in Chapter 7. We also note that any kinds of censored data can be modelled, as long as the corresponding probabilities are correctly specified.

We can readily extend this model to the multivariate case where several kinds of durations are measured. This will require one extra lowest level to be inserted to describe the multivariate structure, with level 2 becoming the between-observation level and level 3 the original level 2. For the logit part of the model we will allow correlations at level 2 where these can be interpreted as point-biserial correlations.

For repeated measures models, where there are different types of duration, we can choose to fit a multivariate model. Alternatively, as discussed in Chapter 4, we may be able to specify a simpler model where the types differ only in terms of a fixed part contribution, or perhaps where there are different variances for each type with a common covariance. As pointed out earlier, we may sometimes wish to treat the first duration length separately and this is readily done by specifying it as a separate response.

Infinite durations

9.12   It is sometimes found that for a proportion of individuals, their duration lengths are extremely long. Thus, some employees remain in the same job for life and some patients may acquire a disease and retain it for the rest of their lives. In studies of social mobility, some individuals will remain in a particular social group for a finite length of time while others may never leave it: such models are sometimes referred to as mover–stayer models. We can treat such durations as if they were infinite. Since any given study will last only for a finite time, it is impossible to distinguish infinite times from those which are right censored. Nevertheless, if we make suitable distributional assumptions we can obtain an estimate of the proportion of infinite survival times.

For a constant θ, given an unobserved duration time, the observation either is right censored with finite duration or has infinite duration so that we replace the probability π_{ij} by $\lambda_{ij} = (1 - \theta)\pi_{ij} + \theta$. In general, θ will depend on explanatory variables and an obvious choice for a model is

$$\text{logit}(\theta_{ij}) = X_{ij}^{(\theta)}\beta^{(\theta)} \tag{9.8}$$

The coefficients in (9.8) may also vary across level 2 units.

Where the observation is not censored we know that it has a finite duration, so that for the infinite duration parameters we have a response variable taking the value zero with a predictor given by $\{1 + \exp - (1 - \theta_{ij})\}^{-1}$. The full model can therefore be specified as a bivariate model where for observed durations we have two responses, one for the uncensored component l_{ij} and one for the parameters $\beta^{(\theta)}$. For the censored observations there is a single response which takes the value one with predictor function

$$\{1 + \exp - [(1 - \theta_{ij})\pi_{ij} + \theta_{ij}]\}^{-1}$$

We can extend the procedures of Chapter 7 to the joint estimation of β, $\beta^{(\theta)}$, noting that for the censored observations, when estimating β, we have

Table 9.2 Log duration of birth interval for Hutterite women. Subscript 1 refers to birth year, 2 to age and 3 to previous death

Parameter	Estimate (s.e.) A	Estimate (s.e.) B	Estimate (s.e.) C
Fixed:			
Intercept	1.97	1.96	1.97
Mother's birth year -1900	-0.021 (0.002)	-0.021 (0.002)	-0.021 (0.002)
Mother's age -20	-0.005 (0.010)	-0.005 (0.010)	-0.005 (0.010)
Previous death	-0.435 (0.079)	-0.436 (0.079)	-0.438 (0.089)
Marriage duration (months)	0.003 (0.001)	0.003 (0.001)	0.003 (0.001)
Random:			
Level 2			
σ_{u0}^2	0.127 (0.017)	0.114 (0.052)	0.121 (0.054)
σ_{u01}		-0.001 (0.002)	-0.001 (0.002)
σ_{u1}^2		0.0001 (0.0001)	0.0001 (0.0001)
σ_{u02}		-0.004 (0.003)	-0.005 (0.003)
σ_{u12}		0.0001 (0.0001)	0.0001 (0.0001)
σ_{u2}^2		0.0005 (0.0003)	0.0006 (0.0003)
Level 1			
σ_{e0}^2	0.549 (0.018)	0.533 (0.018)	0.522 (0.018)
σ_{e3}^2			0.200 (0.108)
-2 log-likelihood		5305.9	5295.5 5290.8

$$\lambda'_{ij}(\beta) = (1 - \theta_{ij})\pi'_{ij}$$

and for estimating $\beta^{(\theta)}$ we have

$$\lambda'_{ij}(\beta^{(\theta)}) = (1 - \pi_{ij})\theta'$$

Examples with birth interval data and children's play episodes

9.13 We first look again at the Hutterite birth interval data. Since all the durations there are uncensored we apply a standard model to the log(birth interval) values. Results are given in Table 9.2.

We see that we can now fit the year of birth and age as random coefficients at level 2. A joint test gives a chi-squared value of 10.4 with 5 degrees of freedom, $P = 0.065$, and they are each separately significant with a significance level of 6%. We have significant heterogeneity at level 1 where the variance within women is greater where there has been a previous death, with a chi squared on 1 degree of freedom of 4.7, $P = 0.03$. As before, mother's birth year and previous death are

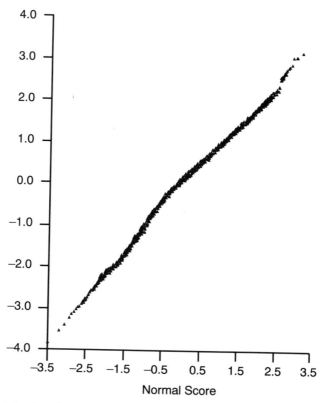

Figure 9.2   Level 1 residuals by Normal scores for Analysis B in Table 9.2.

Surviving fraction

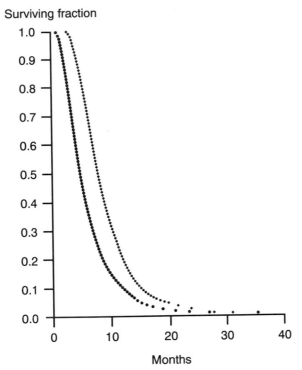

Months

Figure 9.3 Estimated survival functions for women with previous live births (upper) and a previous death; born in 1900, age 20, 12 months marriage.

associated with a decrease, and duration of marriage with an increase, in birth interval. The estimated surviving fraction will, in general, depend on the level 1 distributional assumption. In the present case, as shown in Fig. 9.2, the level 1 standardized residuals show little departure from Normality and Fig. 9.3 shows the estimated surviving fraction based on Normality for women born in 1900, with marriage duration 12 months, aged 20 and with a previous live birth. Figure 9.3 is similar to Fig. 9.1 based on the proportional hazards model. In fact, the two lines actually cross at about 30 months, as a result of the different level 1 variances for those with a previous live birth as opposed to a death.

We now look at some data which exhibit more extensive variance heterogeneity at level 1. They measure the number of days spent by pre-school children either at home or in one of six different kinds of pre-school play activity. For each of 249 children there were up to 12 periods of activity.

The response is the logarithm of the number of days, and covariates are the type of episode, with home chosen as the base category and the education of the mother measured on a 7-point scale ranging from no education beyond minimum school leaving age (0) to university degree (6). Nineteen of the episodes were right censored and 25 were left censored, being less than one day.

The multilevel structure is that of episodes within children. The model is also multivariate with the type of play as six response variables, covarying at the level

Table 9.3 Log duration analysis of children's play episodes: extreme value distribution

Parameter	A (s.e.)	B (s.e.)
Fixed:		
Intercept	2.19	2.18
Play 1	−0.12 (0.11)	−0.13 (0.11)
Play 2	0.20 (0.08)	0.18 (0.08)
Play 3	0.00 (0.13)	0.00 (0.13)
Play 4	0.87 (0.12)	0.95 (0.11)
Play 5	0.28 (0.09)	0.28 (0.09)
Play 6	0.15 (0.09)	0.14 (0.08)
Mother Educn.	−0.05 (0.02)	−0.05 (0.02)
Random		
Level 1:		
Overall	0.75	
Home		0.76
Play 1		1.23
Play 2		0.83
Play 3		0.79
Play 4		0.40
Play 5		0.65
Play 6		0.57

Level 2 covariance matrix. Analysis A (analysis B in brackets)

	Play 1	Play 2	Play 4
Play 1	0.34 (0.0)		
Play 2	0.11 (0.0)	0.20 (0.17)	
Play 4	−0.28 (0.0)	0.13 (0.09)	0.07 (0.23)

of the child. Table 9.3 shows the results of an analysis where there is a single be-tween-child variance and where it is allowed to differ for each type of episode. The between-episode-within-child variance is also allowed to vary for different episodes. The level 1 residuals for the continuous response part of the model show some evidence of non-Normality and we therefore show the results for the extreme value distribution. Because of the relatively small amount of censoring there is little difference for the parameter estimates between analyses making other distributional assumptions.

We see that there is quite substantial variation at both levels. At level 2 there was between-children variation only for play types 1, 2 and 4. A proportional hazards model fitted to these data did not show any between-child variation. In general, the semiparametric proportional hazards model will not detect some of the relation-ships apparent from fitting parametric models, although it has the advantage that it does not make strong distributional assumptions.

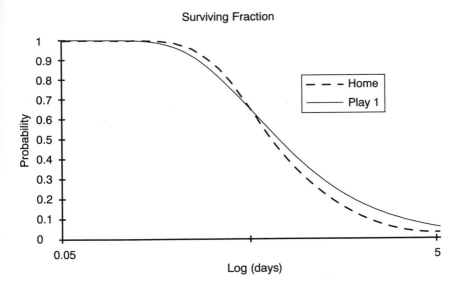

Figure 9.4 Estimated surviving probability of play episodes.

Figure 9.4 shows the predicted probabilities of home and play type 1 episodes lasting beyond various times expressed in log (days). The crossing of the lines is now much clearer as a consequence of the different level 1 variances.

10

Multilevel Models with Measurement Errors

Errors of measurement

10.1 Many measurements are made with substantial error components, especially in the social and biological sciences. If the measurement were to be repeated we would not expect always to get an identical result. In some cases, such as the measurement of individual height or weight, the errors may be so small that they can safely be ignored in practice. In other cases, for example for educational tests and attitude measures, this usually will not be true and a failure to ignore errors may lead to incorrect inferences. Fuller (1987) provides a comprehensive account of methods for dealing with measurement errors in linear models and this chapter extends some of those procedures to the multilevel model. The basic model for measurement errors in a 2-level model for the hth explanatory variable and the response, is as follows

$$
\left.
\begin{aligned}
Y_{ij} &= y_{ij} + q_{ij} \\
X_{hij} &= x_{hij} + m_{hij} \\
\mathrm{cov}(q_{ij} q_{i'j}) &= \mathrm{cov}(m_{hij} m_{hi'j}) = 0
\end{aligned}
\right\}
\tag{10.1}
$$

where upper case letters denote the observed measurements and lower case the underlying 'true' measurements. Thus, we can think of these true measurements as being the expected values of repeated measurements of the same unit where the measurement errors are independent and are also independent of the true values. We define the reliability of the hth explanatory variable as

$$
R_h = \sigma_{hx}^2 / \sigma_{hX}^2 = (\sigma_{hX}^2 - \sigma_{hm}^2) / \sigma_{hX}^2
\tag{10.2}
$$

that is, the variance of the true values divided by the variance of the observed values. This immediately raises two problems. When we measure such things as attitudes or educational achievement, we cannot carry out repeat measurements to obtain estimates of the σ_{hm}^2 because the measurement errors cannot be assumed to be independent. Another way of viewing this is to say that the process of measurement itself has changed the individual being measured, so that the underlying true value has also changed.

The second problem is that we have to define a suitable population. The definition of reliability is population dependent, so that, for example, if the measurement error variance σ_{hm}^2 remains constant but the population heterogeneity of the true values increases then the reliability will increase. Thus, the reliability may be lower within population subgroups, defined by social status say, than in the population as a whole. In particular, the reliability of a test score may be smaller within level 2 units, say schools, than across all students.

In this chapter we shall assume that the variances and covariances of the measurement errors are known, or rather that suitable estimates exist. The topic of measurement error estimation is a complex one, and there are, in general, no simple solutions, except where the assumption of independence of errors on repeated measuring can be made. The common procedure, especially in education, of using 'internal' measures based upon correlational patterns of test or scale items, is unsatisfactory for a number of reasons and may often result in reliability estimates which are too high. Ecob and Goldstein (1983) discussed these and proposed some alternative estimation procedures. McDonald (1985) and other authors discussed the exploration and estimation of measurement error variances within a structural equation model, which has much in common with the suggestions of Ecob and Goldstein (1983). Because estimates of measurement error variance are generally imprecise it is useful to study the effects of varying them and we will illustrate this in examples.

Measurement errors in level 1 variables

10.2 We use a two-level model to show how measurement errors can be incorporated into an analysis. A full derivation is given in Appendix 10.1. We write for the true model

$$y_{ij} = (x\beta)_{ij} + (z_u u)_j + (z_e e)_{ij} \qquad (10.3)$$

where for now we assume that the explanatory variables for the random variables are measured without error, which will be true for variance component models. We assume that it is this true model for which we wish to make estimates. In some situations, for example where we wish simply to make a prediction for a response variable based upon observed values, it is appropriate to treat these values without correcting for measurement errors. If we wish to understand the nature of any underlying relationships, however, we require estimates for the parameters of the true model.

For the observed variables (10.3) gives

$$Y_{ij} = q_{ij} - (m\beta)_{ij} + (X\beta)_{ij} + (z_u u)_j + (z_e e)_{ij} \qquad (10.4)$$

In Appendix 10.1 we show that the fixed effects are estimated by

$$\hat{\beta} = \hat{M}_{xx}^{-1}\hat{M}_{xy}$$

$$\hat{M}_{xx} = X^T\hat{V}^{-1}X - C_{\Omega_1} \left.\vphantom{\begin{matrix}1\\1\\1\end{matrix}}\right\} \qquad (10.5)$$

$$C_{\Omega_1} = \left\{\sum_i \sigma^{ii}\sigma^i_{(h_1,h_2)m}\right\}$$

where $\sigma^i_{(h_1,h_2)m}$ is the covariance between the measurement errors for explanatory variables h_1, h_2 for the ith level 1 unit. The last expression in (10.5) is a correction matrix for the measurement errors and has elements which are weighted averages of the covariances of the measurement errors for each level over all the level 1 units in the sample with the weights being the diagonal elements of V^{-1}. In variance component models this is a simple average over the level 1 units, and in the common case where the covariance matrix of the measurement errors is assumed to be constant over level 1 units we have

$$C_{\Omega_1} = \mathrm{tr}(V^{-1})\Omega_{1m}, \quad \Omega_{1m} = \{\sigma_{(h_1,h_2)m}\} \qquad (10.6)$$

An approximation to the covariance matrix of the estimates is given in Appendix 10.1 as is an expression for the estimation of the random parameters. For the constant measurement error covariance case with no measurement errors in the response variable this covariance matrix is given by

$$\hat{M}_{xx}^{-1}(X^T\hat{V}^{-1}X + X^T\hat{V}^{-2}T_mX)\hat{M}_{xx}^{-1} \left.\vphantom{\begin{matrix}1\\1\end{matrix}}\right\} \qquad (10.7)$$

$$T_{1m} = (\hat{\beta}^T\Omega_{1m}\hat{\beta})I_{(n)}$$

and in the estimation of the random parameters the term T_{1m} is subtracted from $\tilde{Y}\tilde{Y}^T$ at each iteration. It is important in some applications to allow the measurement error variances to vary as a function of explanatory variables. For example, in perinatal studies, the measurement of gestation length may be quite accurate for some pregnancies where careful records are kept but less so in others.

Where the explanatory variables have random coefficients the above results are modified somewhat and the details are given in Appendix 10.1.

Measurement errors in higher level variables

10.3   Where variables are defined at level 2 or above with measurement errors we have analogous results, with details given in Appendix 10.1. Thus, the correction term to be used in addition to C_{Ω_1} with a constant measurement error covariance matrix in a 2-level model is

$$C_{\Omega_2} = \left(\sum_j J_{n_j}^T V_j^{-1} J_{n_j}\right)\Omega_{2m} \qquad (10.8)$$

where J_n is a vector of ones of length n and V_j is the jth block of V.

A case of particular interest is where the level 2 variable is an aggregation of a level 1 variable. Woodhouse *et al* (1995) consider this case in detail and give detailed derivations. Consider the case where we have a level 2 variable which is the mean of a level 1 variable

$$X_{1.j} = \frac{1}{n_j} \sum_i X_{1ij}$$

The variance over the whole sample is therefore given by

$$\text{var}(X_{1.j}) = n_j \frac{\text{var}(X_{1ij})}{n_j^2} + n_j(1-n_j)\frac{\text{cov}(X_{1ij}X_{1i'j})}{n_j^2}$$

$$= \frac{1}{n_j}\sigma_{(1)}^2(X_1) + \frac{n_j-1}{n_j}\sigma_{(2)}^2(X_1)$$

(10.9)

where we assume constant variances and covariances within level 2 units for the X_{1ij}. The number of level 1 units actually measured in the jth level 2 unit is n_j out of a total number of units N_j. Straightforward estimates of the parameters can be obtained by carrying out a variance components analysis with X_{1ij} as the response, fitting only the overall mean in the fixed part, so that the covariance is the level 2 variance estimate.

For the true values we have an analogous result where now we consider the variance of the mean of the true values for *all* the level 1 units in each level 2 unit. There are, in effect, two sources of error in $X_{1.j}$. There is the error inherent in the level 1 measurement X_1 which is averaged across the level 1 units in each level 2 unit and there is the sampling error which occurs when $n_j < N_j$; that is, not all the units in the level 2 unit are measured. Thus, the true value is the average for all the level 1 units in each level 2 unit of the true level 1 measurements. Since the measurement errors are assumed independent we have

$$\text{var}(x_{1.j}) = \frac{1}{N_j}\sigma_{(1)}^2(x_1) + \frac{N_j-1}{N_j}\sigma_{(2)}^2(x_1)$$

(10.10)

This gives us the following expression for the required measurement error variance for the aggregated variable

$$\sigma_{1.m}^2 = \left(\frac{1}{n_j} - \frac{R_1}{N_j}\right)\sigma_{(1)}^2(X_1) - \left(\frac{1}{n_j} - \frac{1}{N_j}\right)\sigma_{(2)}^2(X_1)$$

(10.11)

where the reliability R_1 is estimated from the level 1 variation.

If both the level 1 observed variable and its aggregate are included as explanatory variables then clearly their measurement errors are correlated and the correlation is given by

$$\frac{1-R_1}{n_j}\sigma_{(1)}^2(X_1)$$

In the expressions for the correction matrices, we have considered the separate contributions from levels 1 and 2. Where there is a 'cross-level' correlation between measurement errors as above then we add the level 1 variable to Ω_{2m} using (10.11) for the covariance together with a zero variance. The measurement error variance for the level 1 explanatory variable becomes a component of Ω_{1m}. A detailed derivation of these results is given by Woodhouse *et al* (1995).

Table 10.1 11-year Normalized mathematics score related to 8-year score, gender and social class for different 8-year score level 1 reliabilities; adjusting for measurement errors at level 1 only

Parameter	Estimate (s.e.) A (R_1=1.0)	Estimate (s.e.) B (R_1=0.9)	Estimate (s.e.) C (R_1=0.8)
Fixed:			
Intercept	0.14	0.11	0.08
8 year score	0.095 (0.0037)	0.107 (0.0042)	0.122 (0.0050)
Gender	−0.044 (0.050)	−0.044 (0.050)	−0.043 (0.052)
Non Manual	0.15 (0.06)	0.11 (0.06)	0.06 (0.06)
Random:			
σ_u^2	0.081 (0.023)	0.081 (0.024)	0.082 (0.024)
σ_e^2	0.423 (0.023)	0.374 (0.023)	0.311 (0.025)
Intra-school correlation	0.16	0.18	0.21

A 2-level example with measurement error at both levels

10.4   We use the Junior School Project data reading score at the age of 11 years as our response with the 8-year mathematics score as the predictor, fitting also social class (Non-manual and Manual) and gender. The scores at age 11 have been transformed to have a standard Normal distribution. In addition, we shall allow for measurement errors in both the test scores. There are a total of 728 students in 48 schools in this analysis.

In the original analyses of these data (Mortimore *et al*, 1988) reliabilities are not given, and for the reasons given above are unlikely to be well estimated. For the purpose of our analyses we investigate a range of reliabilities from 0.8 to 1.0 to study the effect of introducing increasing amounts of measurement error.

It can be seen in Table 10.1 that the inferences about the fixed parameters and the level 1 variance and intra-school correlation change markedly in moving from an assumption of zero measurement error to a reliability of 0.8. The increase in the intra-school correlation reflects the fact that it is only the level 1 variance which decreases as the reliability falls. The difference between the children from non-manual and manual backgrounds is considerably reduced as the reliability decreases.

We now look at the effect of additionally adjusting for measurement error in the response variable. To illustrate this we look at the effects on the individual parameters for a range of values for the reliabilities of both response and explanatory variables.

As the response variable reliability decreases, so does the level 1 variance estimate. Likewise, as the reliability of the 8-year score decreases, the level 1 variance decreases. The combined effect of both reliabilities being 0.8 produces a variance which is a quarter of the estimate which assumes no unreliability. When both the reliabilities reach the value of 0.7 the level 1 variance decreases to zero! By contrast, the level 2 variance is hardly altered. For the coefficient of the 8-year score and social class, the greatest change is with the reliability of the 8-year score.

Table 10.2 Parameter estimates (standard errors) for values of explanatory and response variables

Eight year score		Response reliability		
		1.0	0.9	0.8
Eight year score reliability	1.0	0.095	0.095	0.095
	0.9	0.107	0.107	0.107
	0.8	0.122	0.122	0.123
Gender		Response reliability		
		1.0	0.9	0.8
Eight year score reliability	1.0	−0.044	−0.044	−0.043
	0.9	−0.044	−0.043	−0.042
	0.8	−0.043	−0.042	−0.041
Non-manual		Response reliability		
		1.0	0.9	0.8
Eight year score reliability	1.0	0.15	0.15	0.16
	0.9	0.11	0.11	0.12
	0.8	0.06	0.06	0.06
Level 2 variance		Response reliability		
		1.0	0.9	0.8
Eight year score reliability	1.0	0.081	0.080	0.079
	0.9	0.081	0.080	0.079
	0.8	0.082	0.081	0.080
Level 1 variance		Response reliability		
		1.0	0.9	0.8
Eight year score reliability	1.0	0.423	0.325	0.226
	0.9	0.374	0.275	0.177
	0.8	0.311	0.212	0.113

As the reliability decreases so the strength of the relationship with 8-year score increases, while the social class difference decreases substantially. The gender difference is changed very little.

Clearly, the requirement of a positive level 1 variance implies particular lower bounds on the reliabilities and measurement error variances, and underlines the importance of obtaining good estimates of these parameters or at least a range of reasonable estimates. The range of intra-school correlation coefficients, from 16% to 21%, also indicates that we need to take care in interpreting small values of such coefficients without adjusting for measurement error.

Multivariate responses

10.5   To model multivariate data, as discussed in Chapter 4, we specify a dummy (0,1) variable for each response and corresponding interactions with other explanatory variables. Then C_{Ω_1}, C_{Ω_2} in (10.5) and (10.8) are modified so that for each level 1 or level 2 unit, the covariance between measurement errors is set to zero when either of the corresponding dummy variables is zero and likewise for the variances. This is equivalent to specifying the same covariance matrix of measurement errors for each set of explanatory variables corresponding to a response variable, with no covariances across these sets. For the response variables, we likewise specify the separate measurement error variances for each one using the general procedures in Appendix 10.1.

Nonlinear models

10.6   Consider the 2-level model (5.3) in Chapter 5 where there are measurement errors in the explanatory variables for the fixed part of the model. In this case, we can obtain an approximate analysis by using the observed values in the updating formulae and replacing the measurement error covariances in (10.5) by

$$(f'_{(i)})^2 \sigma^i_{(h_1, h_2)m} \qquad (10.12)$$

where $f'_{(i)}$ is the first differential of the nonlinear function for the ith level 1 unit with a corresponding expression for level 2 measurement errors. The derivation of (10.12) is given in Appendix 10.1. Where the variables with measurement errors have random coefficients we likewise replace the corresponding measurement error covariances in Section 10.1.3 of Appendix 10.1 by (10.12).

Measurement errors for discrete explanatory variables

10.7   Assume that we have a categorical explanatory variable with r categories. We shall consider only a single such variable, since multiple variables can, in principle, be handled by considering the p-way table based upon them as a single vector. In practice, it will often be reasonable to assume that their measurement errors are uncorrelated so that they can be considered separately. Likewise we can often assume that measurement errors in discrete explanatory variables are uncorrelated with those in continuous variables. The following derivations parallel those given by Fuller (1987, Section 3.4). We consider only level 1 explanatory variables, but the extension to higher levels follows straightforwardly.

Let $A_{(i)}$ $(1 \times r)$ be a row vector for the ith level 1 unit containing a one for the category which is observed and zeros elsewhere. Let k_{mn} be the probability that a level 1 unit with true category n is observed in category m. We write

$$K = \{k_{mn}\}, \text{ where } K_m \text{ is the } m\text{th column of } K$$

and define

$$X^T_{(i)} = K^{-1} A^T_{(i)} \qquad (10.13)$$

If x_i is the true value we write

$$A_{(i)} = x_{(i)} + \epsilon_{(i)}, \quad E(A_{(i)} \mid x_{(i)}) = x_{(i)} K^T$$

We also write

$$X_{(i)} - x_{(i)} = m_{(i)}$$

so that

$$E(m_{(i)} \mid x_{(i)}) = 0$$

which gives the familiar form for the errors in variables model where the unknown true value $x_{(i)}$ is uncorrelated with the measurement error. The $X_{(i)}$ become the new set of observed values and interest is in the regression on the true category values $x_{(i)}$. The vector $x_{(i)}$ consists of a single value of one and remainder zero. We have

$$\text{cov}(A_{(i)}^T \mid x_{(i)} = l_m) = \sum_{(i)(m)} = \text{diag}(K_m) - K_m K_m^T$$

where l_m is an r-dimensional vector with 1 in the mth position and zeros elsewhere. For the ith level 1 unit define

$$\Omega_{(i)m} = \text{cov}(m_{(i)}^T \mid x_{(i)} = l_m) = K^{-1} \sum_{(i)(m)} K^{-1^T} \tag{10.14}$$

and we use as our estimate of the covariance matrix of measurement errors the matrix in (10.14) conditional on the observed $A_{(i)}$.

$$\hat{\Omega}_{(i)m} = \Omega_{(i)m} \left[\frac{P(x_{(i)} = l_m)}{P(A_{(i)} = l_m)} \right] \tag{10.15}$$

The term in square brackets can be estimated as follows. If μ_A, μ_x are the observed and true vectors of probabilities for the categories, then

$$\mu_x = K^{-1} \mu_A$$

and given the sample estimate of μ_A we can estimate μ_x. The estimate given by (10.15) is then used as in the case of continuous explanatory variables measured with error. In the general model, the number of explanatory variables will generally be one less than the number of categories, with one of the categories chosen as the base and omitted.

In practice, the matrix of probabilities K is normally assumed constant but can itself depend on further explanatory variables. Often we will not have a good estimate of it, and we may need to make some simplifying assumptions. In the case of a binary variable it may be possible to assume equal misclassification probabilities, in which case only a single value needs to be determined, and in practice a range of values can be explored.

APPENDIX 10.1

Measurement errors

The basic 2-level model

10.1.1 We consider the 2-level model and write

$$
\left.
\begin{aligned}
Y_{ij} &= y_{ij} + q_{ij} \\
X_{hij} &= x_{hij} + m_{hij} \\
\mathrm{cov}(q_{ij}q_{i'j}) &= \mathrm{cov}(m_{hij}m_{hi'j}) = 0 \\
E(q_{ij}) &= E(m_{hij}) = 0 \\
\mathrm{cov}(m_{h_1ij}m_{h_2ij}) &= \sigma^i_{(h_1h_2)jm}
\end{aligned}
\right\}
\tag{10.1.1}
$$

for the hth explanatory variable with measurement error vector m_h and with q as the measurement error vector for the response. We use upper case for the observed and lower case for the 'true' values which are the expected values of the observed measurements. Each level 1 unit may have its own set of measurement error variances. Where we have a level 2 explanatory variable, then the measurement error is constant within a level 2 unit.

We write the 'true' model in the general form

$$
y_{ij} = (x\beta)_{ij} + (z_u u)_j + (z_e e)_{ij}
\tag{10.1.2}
$$

which gives the model for the observed variables as

$$
\begin{aligned}
Y_{ij} &= q_{ij} - (m\beta)_{ij} + (X\beta)_{ij} + (z_u u)_j + (z_e e)_{ij} \\
m &= \{m_h\}
\end{aligned}
\tag{10.1.3}
$$

For the true values write

$$
\begin{aligned}
M_{xx} &= x^{\mathrm{T}}V^{-1}x, \quad M_{xy} = x^{\mathrm{T}}V^{-1}y \\
\hat{\beta} &= M_{xx}^{-1}M_{yy}
\end{aligned}
\tag{10.1.4}
$$

Now

$$X^T V^{-1} X = (x + m)^T V^{-1}(x + m)$$
$$= x^T V^{-1} x + m^T V^{-1} x + x^T V^{-1} m + m^T V^{-1} m \qquad (10.1.5)$$

so that

$$E(X^T V^{-1} X) = x^T V^{-1} x + E(m^T V^{-1} m) \qquad (10.1.6)$$

If we further assume that q and m are uncorrelated then we have

$$E(X^T V^{-1} Y) = x^T V^{-1} y \qquad (10.1.7)$$

Thus, to estimate the fixed parameters we require $E(m^T V^{-1} m)$ and we now consider how to obtain this for measurement errors at both level 1 and level 2. We then consider the problem of obtaining estimates of the random parameters required to form V.

Parameter estimation

10.1.2   For errors of measurement in level 1 units the (h_1, h_2) element of $E(m^T V^{-1} m)$ is

$$\left. \begin{array}{c} \displaystyle\sum_{i=1}^{N} \sigma^{ii} \sigma^i_{(h_1 h_2) jm} \\[2ex] \text{with } C_{\Omega_1} = \{\displaystyle\sum_i \sigma^{ii} \sigma^i_{(h_1 h_2) jm}\} \end{array} \right\} \qquad (10.1.8)$$

where N is the total number of level 1 units. In the case where each level 1 unit has the same covariance matrix of measurement errors we have

$$C_{\Omega_1} = \text{tr}(V^{-1})\Omega_{1m}, \quad \Omega_{1m} = \{\sigma_{(h_1 h_2)m}\} \qquad (10.1.9)$$

For errors of measurement in level 2 explanatory variables we have

$$C_{\Omega_2} = \sum_j (J_{(1,n_j)} V_j^{-1} J_{(n_j,1)})\Omega_{2jm} \qquad (10.1.10)$$

where Ω_{2jm} is the covariance matrix of measurement errors for the jth level 2 block, and $J_{(r,s)}$ is an $(r \times s)$ matrix of ones. In Chapter 10 we discuss how to obtain the Ω_{2jm} for level 2 variables which are aggregates of level 1 variables.

For the measurement error corrected estimate of the fixed coefficients we have

$$\hat{M}_{xx} = M_{XX} - C_{\Omega_1} - C_{\Omega_2} \qquad (10.1.11)$$

For the random component based upon the model with observed variables we write the residual $v_{ij} = (z_u u)_j + (z_e e)_{ij} + q_{ij} - (m\beta)_{ij}$, $v = \{v_{ij}\}$ which gives

$$\left. \begin{array}{c} E(vv^T) = V + \displaystyle\bigoplus_{ij} \sigma_{ijq}^2 + T_1 + T_2 \\[2ex] T_1 = \displaystyle\bigoplus_{ij}(\hat{\beta}^T \Omega_{1ijm}\hat{\beta}), \quad T_2 = \displaystyle\bigoplus_j (\hat{\beta}^T \Omega_{2jm}\hat{\beta})J_{(n_j,n_j)} \end{array} \right\} \qquad (10.1.12)$$

where σ_{ijq}^2 is the measurement error variance for the ijth response measurement. Thus, the quantity $\underset{ij}{\oplus}\sigma_{ijq}^2 + T_1 + T_2$ should be subtracted from the sum of products

matrix $\tilde{Y}\tilde{Y}^T$ at each iteration, when estimating the random parameters.

The covariance matrix of the estimated fixed coefficients is given by

$$\hat{M}_{xx}^{-1}(X^T V^{-1} X + X^T V^{-1} Q V^{-1} X + X^T V^{-2}[T_1 + T_2] X) \hat{M}_{xx}^{-1}$$

$$Q = \underset{ij}{\oplus}\sigma_{ijq}^2 \qquad (10.1.13)$$

This expression ignores any variation in the estimation of the measurement error variance itself, although Goldstein (1986) includes terms for this.

Random coefficients for explanatory variables measured with error

10.1.3   We have assumed so far that the coefficients of variables with measurement error are not random. Where such coefficients are random the above formulae are modified as follows.

From (10.1.1) and (10.1.2) we have

$$E[\tilde{Y}\tilde{Y}^T] = E[(z_u u + z_e e)(z_u u + z_e e)^T] + E(qq^T) + E[(m\beta)(m\beta)^T] \qquad (10.1.14)$$

$$\tilde{Y}\tilde{Y}^T = (Y - X\beta)(Y - X\beta)^T$$

If Z refers to the observed values of the explanatory variables for the random variables then, if we assume that all the measurement errors are independent of all the true values, we obtain for measurement errors of level 1 variables

$$E[(Z_u u + Z_e e)(Z_u u + Z_e e)^T] = E[(z_u u + z_e e)(z_u u + z_e e)^T]$$
$$+ E[(mu)(mu)^T + (me)(me)^T]$$

where

$$E[(me)(me)^T] = S_{me} = \underset{ij}{\oplus}S_{(ij)me}$$

$$S_{(ij)me} = \sum_{h=1}^{p}\sigma_{(ij)hm}^2\sigma_{(ij)he}^2 + 2\sum_{h_1<h_2}^{p}\sigma_{(ij)h_1h_2m}\sigma_{(ij)h_1h_2e}$$

$$(10.1.15)$$

where $\sigma_{(ij)h_1h_2m}$, $\sigma_{(ij)h_1h_2e}$ are the covariances for the h_1, h_2 measurement errors and random coefficients for the ijth level 1 unit respectively, and with a similar expression for $E[(mu)(mu)^T] = S_{mu}$. Where we have measurement errors at level 2 then for the level 1 random variables the expression (10.1.15) holds. Where the random variables are at level 2 we have

$$S_{(j)mu} = \sum_{h=1}^{p}\sigma_{(j)hm}^2\sigma_{(j)hu}^2 J_{(n_j,n_j)} + 2\sum_{h_1<h_2}^{p}\sigma_{(j)h_1h_2m}\sigma_{(j)h_1h_2u}J_{(n_j,n_j)}$$

$$S_{mu} = \underset{j}{\oplus}S_{(j)mu} \qquad (10.1.16)$$

For estimating the random parameters we form the corrected sum of products matrix

$$\tilde{Y}\tilde{Y}^{\mathrm{T}} - Q - T_1 - T_2 + S_{me} + S_{mu} \tag{10.1.17}$$

and carry out the generalized least squares regression of this on the random parameter design matrix using the observed explanatory variables and the value of V based on these observed values.

For estimating the fixed part coefficients we use the value of V based upon the estimated true values of the random parameters with the observed explanatory variables less the correction term

$$S_{me} + S_{mu}$$

Note that when carrying out estimations for missing data (Chapter 11) this correction term is added to V.

Nonlinear models

10.1.4   Consider first the case where just the fixed part explanatory variables have measurement errors at level 1 in the single component 2-level nonlinear model for the ith level 1 unit

$$y_{(i)} = f_{(i)}(X\beta + \text{random})$$

which yields the linearization

$$y_{(i)} - \{f_{(i)}(X\beta) - \sum_k \beta_{k,t} x_{(i)k}^*\} = \sum_k \beta_{k,t+1} x_{(i)k}^* + \text{random terms} \tag{10.1.18}$$

where the explanatory variables are the observed measurements and the coefficients are the required ones corrected for measurement error and $x_{(i)k}^* = f_{(i)}' x_{(i)k}$. Consider the expansion of $f_{(i)}$ for the measurement error terms, to a first-order approximation,

$$f_{(i)} = f_{(i),m_k=0} + \sum_k f_{(i),u_k=0}' m_{(i)k} \beta_{k,t} \tag{10.1.19}$$

Thus, we can use the observed explanatory variables with measurement error as an approximation to the use of the true values in the updating formulae, with $(f_{(i)}')^2 \sigma_{(h_1,h_2)m}^i$ replacing $\sigma_{(h_1,h_2)m}^i$ in (10.5). Where the variables with measurement errors have random coefficients, we likewise replace the corresponding measurement error covariances in the previous Section 10.1.3 by the same expressions.

11

Software for Multilevel Modelling; Missing Data and Multilevel Structural Equation Models

Software for multilevel analysis

11.1 Traditionally, statistical analysis packages for the analysis of linear or generalized linear models have assumed a single level model, and generally also a single random variable. For the models described in this book such software packages are clearly inadequate, and this led, in the mid 1980s, to the development of four special purpose packages for fitting multilevel models. One of these, GENMOD (Mason *et al*, 1988), is no longer generally available. The other three are HLM (Bryk *et al*, 1988), ML3 (Prosser *et al*, 1991) and VARCL (Longford, 1988). A comparative detailed review of these four packages has been carried out by Kreft *et al* (1994). In their original form HLM, ML3 and VARCL were designed for continuous Normally distributed response variables and all three produced maximum likelihood (ML) or restricted maximum likelihood (REML) estimates. All three were soon able to fit 3-level models and VARCL and ML3 developed procedures for fitting binomial and Poisson response models using the first-order marginal approximation described in Chapter 5. In addition, VARCL is able to fit a variance components model with up to nine levels. Subsequently, the major statistical packages, notably BMDP, SAS and GENSTAT, have begun to incorporate procedures for ML and REML estimation for Normal response models. The packages EGRET and SABRE will obtain ML estimates for a 2-level logit response model. A Bayesian package using Gibbs Sampling, BUGS, is also available. Appendix 11.1 contains details of where these and other programs can be obtained.

The two packages, ML3 and BUGS, are able to fit nearly all the models described in this book, although not currently structural equation models. These latter models can be fitted by the program BIRAM, listed in Appendix 11.1. The successor to ML3, Mln allows an effectively unlimited number of levels to be fitted, together with case weights, measurement errors and robust estimates of standard errors. It also has a high level MACRO language which will allow a wide range of special purpose facilities to be incorporated. Some of the papers referenced in earlier chapters have carried out their estimation procedures using

special purpose software written in statistical programming languages such as S
Plus or Gauss. For the most part, however, this approach is computationall
inefficient for the analysis of large and complex data sets, and the use of one of th
special purpose packages is then essential, even when powerful mainfram
computers are used. The general purpose packages, SAS, GENSTAT and also ML
allow a wide variety of data manipulations to be carried out within the software
whereas the others tend to demand a somewhat rigid data format with limite
possibilities for data transformations etc.

It is reasonable to expect that the standard multilevel models will soon b
available within most of the major general purpose statistical packages. For the
more complex models, such as those with multivariate outcomes, nonlinea
relationships and complex variation at all levels, it will be important to have a use
interface which assists understanding the complexity of structure when specifying
models and when interpreting output. Because the level of complexity of multileve
models is greater than that associated with single level linear or generalized linea
models, the importance of helpful user interfaces cannot be overemphasized if the
best use is to be made of these models. The ability to work interactively in a
graphical environment will also be important and it will be necessary for programs
to optimize computations so that very large and complex datasets can be handled
within a reasonable time (Goldstein and Rasbash, 1992).

Design issues

11.2   When designing a study where the multilevel nested structure of a
population is to be modelled, the allocation of level 1 units among level 2 units and
the allocation of these among level 3 units etc, will clearly affect the precision of
the resulting estimates of both the fixed and random parameters. The situation
becomes more complex when there are random cross classifications and where
there are several random coefficients. There are generally differential costs
associated with sampling more level 1 units within an existing level 2 unit as
opposed to selecting further level 1 units in a new level 2 unit. At the present time
there appears to be little empirical or theoretical work on issues of optimum design
for multilevel models.

Some approximations for studying the standard errors of the fixed coefficients
have been derived by Snijders and Bosker (1993) in the case of a simple 2-level
variance components model. They are concerned with students sampled within
schools and assume that the cost of selecting a student in a new school is a fixed
constant times the cost of selecting a student in an already selected school. They
also assume that there is a minimum of 11 students per school. They tend to find
that, where this constant is greater than 1 and the total number of students to be
sampled is fixed, the sample of schools should be as large as possible, although this
will not necessarily be true for all the coefficients of interest.

Where cost information is available, together with some idea of parameter
values, perhaps from a pilot study, then a guide to design efficiency can be obtained
by simulating the effect of different design strategies and studying the resulting
characteristics of the parameter estimates, such as their mean squared errors. This
will be time consuming, however, since for each design a number of simulated
samples will be required. On the other hand, in certain areas, such as that of school

effectiveness or animal and human growth studies, where information about costs and parameter values is often available, it would be possible to derive some generally useful results.

Missing data

11.3 A characteristic of most large scale studies is that some of the intended measurements are unavailable. In surveys, for example, this may occur through chance or because certain questions are unanswered by particular groups of respondents. We are concerned with missing values of explanatory variables in a multilevel model. An important distinction is made between situations where the existence of a missing data item can be considered a random event and where it is informative and the result of a non-random mechanism. Randomly missing data may be missing 'completely at random' or 'at random' conditionally on the values of other measurements. The following exposition will be concerned with these two types of random event. Where data cannot be assumed to be missing at random, one approach is to attempt to model the 'missingness' mechanism, and then to predict values from this model. Such predictions can be treated in similar ways to those described below.

We consider the problem of missing data in two parts. First we develop a procedure for predicting data values which are missing and then we study ways of obtaining model parameter estimates from the resulting 'filled-in' or 'completed' data set. The prediction will use those measurements which are available, so that data values which are missing at random conditional on these measurements can be incorporated. Detailed discussions of missing data procedures are given by Rubin (1987) and Little (1992).

The basic exposition will be in terms of a single level model for simplicity, pointing out the extensions for multilevel models.

Creating a completed data set

11.4 Consider the ordinary linear model

$$y_i = \beta_0 + \beta_1 x_{1i} + \beta_2 x_{2i} + e_i \tag{11.1}$$

for the ith unit in a single level model. Suppose that some of the x_{1i} are missing completely at random (MCAR) or conditionally missing at random (MAR) conditional on X_2. Label these unknown values x_{1i}^*. We consider the estimation of these by predicting them from the remaining observations and the parameter set θ for the prediction model, namely

$$\hat{x}_{1i}^* = E(x_{1i}^* \mid x_{2i}, y_i, \theta) \tag{11.2}$$

Where we have multivariate Normal data, the prediction (11.2) is simply the linear regression of X_1 on X_2, Y, where the coefficients of this regression prediction are obtained from efficient (for example maximum likelihood) estimates of the parameters of the multivariate Normal distribution. This can be achieved efficiently using the procedures for modelling multivariate data described in Chapter 4. We shall consider the case of non-Normal data later.

We define a multivariate model with three response variables, Y, X_1, X_2 and three corresponding dummy variables, say Z_0, Z_1, Z_2. Some level 2 units will have all three response variables, but others will have only two where X_1 is missing. Write this as the 2-level model

$$v_{ij} = \beta_{0j}z_{0ij} + \beta_{1j}z_{1ij} + \beta_{2j}z_{2ij}$$

$$\beta_{0j} \sim N(\mu_Y, \sigma_Y^2), \quad \beta_{hj} \sim N(\mu_h, \sigma_h^2), \quad h = 1, 2 \tag{11.3}$$

together with the three covariances to give the (2×2) covariance matrix Ω_{XX} and covariance vector Ω_{XY}. This model will produce efficient (ML in the Normal case) estimates of the parameters in (11.1)

$$\hat{\theta} = \Omega_{XX}^{-1}\Omega_{XY} \tag{11.4}$$

Thus, for any missing value we can use the parameters from (11.4) to predict X_1 from X_2, Y_3. These predicted values are just the estimated level 2 residuals from (11.3) for the missing values. Clearly this procedure extends to any number of variables with any pattern of missing data. We simply formulate the model as a multivariate response by introducing dummy variables for each variable and then estimating the residuals for the resulting 2-level model and choosing the appropriate residuals to fill in the missing values. This procedure extends in a straightforward way to multilevel data.

Suppose we have a two-level data set with some explanatory variables measured at level 1 and some at level 2 and various values missing. We specify a 3-level multivariate response model where some of the responses are at level 2 and some at level 3. At level 2 of this model we estimate a covariance matrix for the original level 1 variables and at level 3 we estimate a covariance matrix for all the variables. For the original level 2 variables with missing values we estimate the residuals at level 3 and use these to fill in missing values. For the original level 1 variables we add the level 3 and the level 2 residuals together to obtain filled-in values.

If we were to use the completed data sets in the usual way to fit a multilevel model the resulting estimates would be biased because the filled-in data are shrunken and have less variation than the original measurements. Little (1992) discussed this problem and in the next section we outline procedures for dealing with it.

Multiple imputation and error corrections

11.5   The usual multiple imputation (Rubin, 1987) procedure proceeds as follows. The predicted values are adjusted to have their correct, on average, distributional properties by sampling from the multivariate distribution of the predicted values. Where we have, as in the above example, just one variable with missing values in a single level Normal model this involves a series of random values chosen from the Normal distribution, with mean the residual estimate $\hat{x}_{1i}^*$ and variance given by the estimated (comparative) variance of this residual estimate. For small samples we should also take account of the sampling variation of the estimated parameters, for example using a bootstrap procedure (Chapter 3).

Where the residuals from two different levels are combined, as described above, several level 1 units within the same level 2 unit share the same level 2 residual so that we will need to sample from the multivariate distribution where the variances are simply the sums of the variances from the two levels and the common covariance is the variance of the level 2 estimate. Where there are several variables with filled-in values then we need to sample from an extended multivariate distribution.

Having generated these 'corrections' we then fit our multilevel model in the usual way and obtain parameter estimates. This process is repeated a number of times, and the final estimates are suitably chosen averages of these sets of estimates. These final estimates are asymptotically efficient with consistent standard errors.

This kind of multiple imputation, in practice, has certain drawbacks. The principal one is the amount of computation required to carry out several analyses, especially in its use with secondary data where different analysts, often with limited resources, wish to work on the same data set. As an alternative, the following procedure is proposed.

For our simple example the imputation procedure implicitly assumes a model of the form

$$x_{1i} = \hat{x}_{1i}^* + w_{1i} \tag{11.5}$$

where the w_{1i} have the variances and covariances for the residuals estimated as above, and zero means. This model is similar to the basic model (10.1) in Chapter 10 for errors of measurement, except that the role of x_{1i} is now that of the 'true' value, which is unknown. If we assume that the two terms on the right-hand side of (11.5) are uncorrelated, then we have

$$\text{var}(x_1) = \text{var}(\hat{x}_1^*) + \text{var}(w) \tag{11.6}$$

We see therefore that to obtain estimates for the fixed coefficients based upon the true values we can apply the same procedures as in the measurement error case but with measurement error variances *added* rather than subtracted from the relevant quantities. Thus, for a 2-level model we have the following, which correspond to (11.5) for a model with p explanatory variables with missing data at level 1. We form

$$
\left.
\begin{aligned}
\hat{M}_{xx} &= X^{*T}V^{-1}X^* + C_{\Omega_1} + C_{\Omega_2} \\
C_{\Omega_1} &= \left\{ \sum_i \sigma^{ij}\sigma^{ij}_{e(h_1,h_2)w} \right\} \\
C_{\Omega_2} &= \sum_j \{ J^{*T}_{n_j(h_1 h_2)} V_j^{-1} J^*_{n_j(h_1 h_2)} \sigma^j_{u_j(h_1 h_2)} \}
\end{aligned}
\right\} \tag{11.7}
$$

substituting sample estimates. For the ijth level 1 unit σ^{ij} is the diagonal term of V^{-1} and $\sigma^{ij}_{e(h_1,h_2)w}$ is the corresponding covariance (or variance) between the (level 1) residuals for variables h_1, h_2 where these are both missing. The vector $J_{n_j(h_1 h_2)}$ contains a one if, for the jth second level unit, variables h_1, h_2 are both missing and zero otherwise. The term $\sigma^j_{u_j(h_1 h_2)}$ is the estimated covariance (or variance) between the (level 2) residuals for variables h_1, h_2.

The estimates of the fixed coefficients are given by

$$\hat{\beta} = \hat{M}_{xx}^{-1}\hat{M}_{xy}$$

The extensions for level 2 explanatory variables and discrete variables (see below) are likewise analogous to those described in Chapter 10.

In the single level case for a single explanatory variable with missing data, these results reduce to the following. Order the completed data so that the imputed observations are grouped together first. Then, ignoring the correction for sampling variation, the adjustment is obtained by replacing (X^TX) by

$$(X^TX) + \begin{pmatrix} n_1\hat{\sigma}_w^2 & \\ 0 & 0 \end{pmatrix} \qquad (11.8)$$

where there are n_1 imputed values. This is very similar to the correction described by Beale and Little (1975), although these authors use an estimate based upon the observed residuals calculated from the complete data cases and approximate the covariance matrix by $\hat{M}_{xx}^{-1}$.

Discrete variables with missing data

11.6   Suppose we have one or more categorical explanatory variables as well as continuous variables with missing values. The first stage procedure is to obtain the predicted values. We can do this by treating all the variables together as a multivariate model with mixed continuous and discrete responses as described in Chapter 7. For each categorical variable we obtain the predicted probabilities of belonging to each category, corresponding to each dummy variable used in the subsequent analysis. For a single level model these would be substituted to form the completed data set. For a 2-level model we would add the level 3 residual from the initial multivariate model to each prediction. Thus, where the categorical variable is at level 1 then for each level 1 unit all the dummy variable values are replaced by estimates. We can obtain the $\sigma_{e(h_1,h_2)w}^{ij}$ together with covariances between discrete and continuous variables from the model estimates (Chapter 7) and the relevant higher level variances and covariances are added to models with further levels. Care is needed with such linear predictions for discrete data and further research is required.

An example with missing data

11.7   We use the Junior School Project data set and model A of Table 10.1 to illustrate the missing data procedure. We have omitted, at random, 15% of the values of the 8-year maths score. Three analyses have been carried out. The first simply omits all the level 1 units with a missing value. The second carries out only the first stage of the analysis to provide a completed data set and then proceeds in the usual way. The third analysis carries out the full missing data procedure.

The first stage consists of estimating the level 2 and level 3 covariance matrices for the response and three explanatory variables (excluding the intercept) and estimating the residuals.

Table 11.1 JSP Mathematics data. Model A is full data analysis, model B omits cases with missing data, model C uses completed data, model D uses full missing data procedure

Parameter	Estimate (s.e.) A	Estimate (s.e.) B	Estimate (s.e.) C	Estimate (s.e.) D
Fixed:				
Constant	0.14	0.12	0.097	0.12
8-year score	0.095 (0.0037)	0.100 (0.0040)	0.105 (0.0037)	0.097 (0.0039)
Gender (boys– girls)	−0.044 (0.050)	−0.087 (0.054)	−0.067 (0.047)	−0.066 (0.051)
Social class (Non Man–Manual)	0.154 (0.057)	0.113 (0.060)	0.107 (0.054)	0.135 (0.058)
Random:				
Level 2				
σ_{u0}^2	0.081 (0.023)	0.083 (0.025)	0.077 (0.022)	0.077 (0.023)
Level 1				
σ_{e0}^2	0.423 (0.023)	0.415 (0.024)	0.378 (0.021)	0.412 (0.023)

We see that in the analysis which retains only the complete cases the standard errors are raised. The analysis which uses the completed data set without adjusting for the uncertainty of the predicted values, tends to underestimate the level 1 variance and also changes the fixed parameter estimates markedly. The corrected analysis using the full missing data procedure tends to give standard errors which are somewhat smaller than the analysis which simply omits level 1 units with missing data.

Multilevel structural equation models

11.8   The theory and application of single level structural equation models, including the special cases of observed variable path models and factor analysis models, is well known (Joreskog and Sorbom, 1979, McDonald, 1985). In this chapter we look at multilevel generalizations of these models. We shall not give details of estimation procedures, which are set out by Goldstein and McDonald (1987), McDonald and Goldstein (1988) with elaborations by Muthen (1989) and Longford and Muthen (1992). McDonald (1994) presents an informal overview.

Consider first a basic 2-level factor model where we have a set of measurements on each student within a sample of schools, together with a set of measurements at the school level which may be aggregated student level measurements. The response measurements of interest whose structure we wish to explore are assumed to be random variables, Normally distributed. A further set of covariates, for example gender or social class, are explanatory variables which we may wish to condition on. For the p level 1 responses we first write a multivariate model with p responses where, in general, some may be randomly missing.

$$y_{hij} = (X\beta)_{hij} + \sum_h e_{hij} z_{hij} + \sum_h u_{hj} z_{hij}$$

This is a 3-level model as described in Chapter 4 with dummy variables for each response with random coefficients at level 2 and level 3. Note that at level 3 (between schools) some of the responses may not vary. Note also that, in general, some of the coefficients of the covariates may vary at level 3 and these would be incorporated as further level 3 random variables along with those above. Reverting to the original 2-level model we now have a set of level 1 random variables e_{hij} and a set of level 2 random variables u_{hj}. A general factor structure for the level 1 variables may involve factors defined at both level 1 and level 2, where we can write

$$e_{hij} = \sum_g \lambda_{1gh} f_{gij}^{(1)} + w_{hij}$$

$$u_{hj} = \sum_g \lambda_{2gh} f_{gj}^{(2)} + w_{hj}$$

for the factor structures at each level, using standard notation. We may wish to identify some of these factors as the 'same' factors at each level, for example by constraining certain loadings to be zero. In general, of course, we may have different random variables at level 1 and level 2, since, for example, some of the variables which vary between students may not vary across schools and vice versa. Thus, we may have an attitude score with no between-school variation and any aggregate level variables by definition will not vary between pupils. The latter, nevertheless, may enter the model with the level 1 random variables as responses, by being part of the level 2 factor structure and contributing to the prediction of the u_{hj} in the above equation. Thus, we can, in principle, consider any level 2 random variables including random coefficients of covariates when modelling the factor structure at this level.

A straightforward and consistent procedure for estimating the parameters of this factor model is to do it in two stages. The first stage involves the estimation of the separate level 1 and level 2 residual covariance matrices as described above using the procedures given in Chapter 4. The second stage involves the factor analysis of these separate matrices using any standard procedure, as described for example by Joreskog and Sorbom (1979) or McDonald (1985). This also automatically deals with any missing responses at either level. McDonald (1993) gives details for maximum likelihood estimators in this case.

The two-stage procedure should be reasonably efficient except where the data are unbalanced, with highly variable numbers of level 1 units within level 2 units. It has the advantage that it can be used for quite general structures. Thus, it extends straightforwardly to any number of hierarchical levels. Furthermore, we can also fit models where there are random cross classifications using the procedures described in Chapter 8. Thus, if students are classified by the primary and the secondary school they attended we can estimate the covariance matrices for level 1 and for both classifications at level 2 and then carry out three separate factor analyses of these matrices.

This procedure also allows us to fit general unconditional path models, with or without latent variables, since the covariance matrices at each level are sufficient for these models. A simple example of such a model without latent variables is as follows

$$y_{ij}^{(1)} = \alpha_1 + \beta_1 x_{ij}^{(1)} + u_j^{(1)} + e_{ij}^{(1)}$$

$$y_{ij}^{(2)} = \alpha_2 + \beta_2 y_{ij}^{(1)} + u_j^{(2)} + e_{ij}^{(2)}$$

where the $y_{ij}^{(1)}$ is regarded as a random variable in both equations. The traditional path model treats $y_{ij}^{(1)}$ in the second of these equations conditionally, so that it can be treated straightforwardly as a bivariate 2-level model. A choice between these two models will depend on substantive considerations, especially where there is a temporal ordering of variables when the conditional model would seem to be more appropriate in general. McDonald (1985) gave an account of estimation for unconditional path models.

A factor analysis example using science test scores

11.9   We use the science data analysed in Chapter 4 to fit a 2-level factor model, to the results in Table 4.4. The factor model is fitted to the estimated residual covariance matrices of this table, omitting the variable Earth Science core. We use first the level 1 and level 2 covariance matrices and fit two models. The first assumes one factor at each level with the loadings constrained to be the same and the second allows the loadings to be different. A model with two factors with loadings constrained to be equal at each level was also fitted but yielded a very high correlation (0.95) between the factors at level 1 and an estimated correlation at level 2 of 1.80! The model where the loading constraints were removed failed to converge. The program BIRAM was used with the solution scaled so that the factor variance equals one (McDonald, 1994). The goodness of fit chi-squared values are approximate, based upon the assumption of equal numbers of level 1 units per level 2 unit.

The unconstrained solution shows a greatly improved fit over the constrained solution. At level 1 both the loadings for the Physics tests are somewhat higher than for the Biology tests with R3 having a much lower correlation with the factor. At school level there is no such clear separation between the loadings.

Future developments

11.10   A wide range of topics has been covered in this volume. Normal response models are well understood and have found many successful applications. Binary response models likewise are finding numerous applications. In the former case, there are now efficient algorithms for fitting multilevel and cross classified models

Table 11.2 Factor analysis of residual covariances of Science achievement data

Variable	Unconstrained loadings (s.e.)		Constrained loadings (s.e.)
	Level 2	Level 1	
Biology core	1.02 (0.01)	0.58 (0.02)	0.61 (0.02)
Biology R3	0.97 (0.08)	0.23 (0.02)	0.26 (0.02)
Biology R4	0.73 (0.05)	0.50 (0.02)	0.52 (0.02)
Physics core	0.96 (0.01)	0.64 (0.02)	0.66 (0.02)
Physics R2	0.87 (0.03)	0.64 (0.02)	0.65 (0.02)
χ^2(d.f)	91.9 (10)		236.5 (15)

with many levels and ways of classification. Likewise, the nonlinear modelling of variance functions, including time series analysis, promises to open up interesting new areas of application.

With anticipated increases in the power of computer hardware, the analysis of very large datasets, including for example population censuses, should become feasible. In the case of binary data, as well as count and multicategory response data and nonlinear models more generally, there is more research required on the properties of different estimators. More simulation studies would be useful here. Bayesian methods such as Gibbs Sampling show considerable promise.

The ability to handle measurement errors and missing data efficiently is important and is a generally neglected area in applied research which tends to ignore measurement errors and treat missing data by omitting complete units. The procedures discussed here will benefit from further development and exploration and this should be a priority area for further research, affecting as it does both consistency and efficiency. Likewise, the issue of design efficiency has hardly been explored at all although it is an important topic practically.

Finally, we have presented a succession of models in previous chapters, dealing separately with each one. We have said little about combinations of these to produce more complex models. For example, we can combine a mixed binary and continuous response model with higher level cross classifications and measurement errors. With models of such complexity both the model specification and interpretation will need to be dealt with carefully. This will be helped by the use of powerful graphical procedures for diagnosis and presentation of model structures, and this is an important area for further development.

APPENDIX 11.1

Addresses for multilevel software packages

BIRAM is available from:
Professor R.P. McDonald
Department of Psychology
University of Illinois
603 E. Danial St.
Champaign, IL 61820
USA

BMDP is available from:
BMDP Statistical Software Inc
1440 Sepulveda Blvd. Suite 316
Los Angeles, CA 90025
USA

BUGS is available from:
MRC Biostatistics Unit
Institute of Public Health
Robinson Way
Cambridge CB2 2SR
UK

EGRET is available from:
Statistics and Epidemiology
Research Corporation
909 Northeast 43 Street, Suite 202
Seattle, WA 98105
USA

GENSTAT is available from:
NAG Ltd
Wilkinson House
Jordan Hill Road
Oxford OX2 8DR
UK

HLM is available from:
Scientific Software Inc
1525 East 53rd St.
Suite 906
Chicago IL 60615
USA

ML3 and Mln are available
from:
Multilevel Models Project
Institute of Education
20 Bedford Way
London WC1H 0AL
UK

ML3, HLM and VARCL
are also available from:
ProGamma
P.O.B. Groningen
The Netherlands

SABRE is available from:
Centre for Applied Statistics
University of Lancaster
Lancaster LA1 4YF
UK

SAS is available from:
SAS Institute Inc
SAS Campus Drive
Cary, NC 27513
USA

References

Aitkin, M. and Longford, N. 1986: Statistical modelling in school effectiveness studies (with discussion). *Journal of the Royal Statistical Society*, A 149, 1–43.

Aitkin, M., Anderson, D. and Hinde, J. 1981: Statistical modelling of data on teaching styles (with discussion). *Journal of the Royal Statistical Society*, A 144, 148–61.

Aitkin, M., Anderson, D., Francis, B. and Hinde, J. 1989: *Statistical modelling in GLIM*. Oxford: Clarendon Press.

Beale, E.M.L. and Little, R.J.A. 1975: Missing values in multivariate analysis. *Journal of the Royal Statistical Society*, B 37, 129–45.

Bennett, N. 1976: *Teaching styles and pupil progress*. London: Open Books.

Bock, R.D. 1992: Structural and nonstructural analysis of multiphasic growth. Chicago: University of Chicago (unpublished).

Breslow, N.E. and Clayton, D.G. 1993: Approximate inference in generalised linear mixed models. *Journal of the American Statistical Association* 88, 9–25.

Bryk, A.S. and Raudenbush, S.W. 1992: *Hierarchical linear models*. Newbury Park: Sage.

Bryk, A.S., Raudenbush, S.W., Seltzer, M. and Congdon, R. 1988: *An introduction to HLM: Computer program and user's guide (2nd Edn)*. Chicago: University of Chicago Dept. of Education.

Burdick, R.K. and Graybill, F.A. 1988: The present status of confidence interval estimation on variance components in balanced and unbalanced random models. *Communications in Statistics: theory and methods* 17, 1165–95.

Burstein, L., Fischer, K.H. and Miller, M.D. 1980: The multilevel effects of background on science achievement: a cross national comparison. *Sociology of Education* 53, 215–25.

Clayton, D.G. 1988: The analysis of event history data: a review of progress and outstanding problems. *Statistics in Medicine* 7, 819–41.

Clayton, D.G. 1991: A Monte Carlo method for Bayesian inference in frailty models. *Biometrics* 47, 467–85.

Clayton, D.G. 1992: Bayesian analysis of frailty models. Cambridge, MRG Biostatistics Unit (unpublished).

Clayton, D.G. and Kaldor, J. 1987: Empirical Bayes estimates of age-standardised relative risks for use in disease mapping. *Biometrics* 43, 671–81.

Cochran, W.G. 1983: *Planning and analysis of observational studies.* New York: Wiley.

Cook, R.D. and Weisberg, S. 1982: *Residuals and influence in regression.* London: Chapman and Hall.

Cox, D.R. 1972: Regression models and life tables (with discussion). *Journal of the Royal Statistical Society,* B 34, 187–220.

Cox, D.R. and Oakes, D. 1984: *Analysis of survival data.* London: Chapman and Hall.

Creswell, M. 1991: A multilevel Bivariate Model. In: Prosser, R., Rasbash, J. and Goldstein, H. *Data analysis with ML3.* London: Institute of Education.

Cronbach, L.J. and Webb, N. 1975: Between class and within class effects in a repeated aptitude × treatment interaction: reanalysis of a study by G.L. Anderson. *Journal of Educational Psychology* 67, 717–24.

Demirjian, A., La Palme, L. and Thibault, H.W. 1982: La croissance staturo-ponderale des enfants Canadien-Français de la naissance à 36 mois. *Union Medicale du Canada.* 112, 153–63.

Derbyshire, M.E. 1987: Statistical rationale for grant-related expenditure assessment (GREA) concerning personal social services. *Journal of the Royal Statistical Society,* A 150, 309–33.

Ecob, R. and Goldstein, H. 1983: Instrumental variable methods for the estimation of test score reliability. *Journal of Educational Statistics* 8, 223–41.

Efron B. 1988: Logistic regression, survival analysis, and the Kaplan–Meier curve. *Journal of the American Statistical Association* 83, 414–25.

Efron B. and Gong, G. 1983: A leisurely look at the Bootstrap, the Jacknife and Cross-validation. *The American Statistician* 37, 36–48.

Egger, P.J. 1992: Event history analysis: discrete-time models including unobserved heterogeneity, with applications to birth history data. University of Southampton, PhD thesis.

Fuller, W.A. 1987: *Measurement error models.* New York: Wiley.

Garrett, M., Fitzmaurice, M. and Laird, N. 1993: A likelihood based method for analysing longitudinal binary responses. *Biometrika* 80, 141–51.

Gilks, W.R., Clayton, D.G., Spiegelhalter, D.J., Best, N.G., McNiel, A.J., Sharples, L.D. and Kirby, A.J. 1993: Modelling complexity: applications of Gibbs Sampling in medicine. (With discussion). *Journal of the Royal Statistical Society,* B 55, 39–102.

Goldstein, H. 1976: Smoking in pregnancy: some notes on the statistical controversy. *British Journal of Preventive and Social Medicine* 31, 13–17.

Goldstein, H. 1979: *The design and analysis of longitudinal studies.* London: Academic Press.

Goldstein, H. 1986: Multilevel mixed linear model analysis using iterative generalised least squares. *Biometrika* 73, 43–56.

Goldstein, H. 1987a: Multilevel covariance component models. *Biometrika* 74, 430–1.

Goldstein, H. 1987b: *Multilevel models in educational and social research.*

London: Griffin.

Goldstein, H. 1987c: The choice of constraints in correspondence analysis. *Psychometrika* 52, 207–15.

Goldstein, H. 1989a: Restricted unbiased iterative generalised least squares estimation. *Biometrika* 76, 622–3.

Goldstein, H. 1989b: Efficient prediction models for adult height. In J.M.Tanner (ed.) *Auxology 88; Perspectives in the science of growth and development.* London: Smith Gordon.

Goldstein, H. 1991: Nonlinear multilevel models with an application to discrete response data. *Biometrika* 78, 45–51.

Goldstein, H. 1992: Statistical information and the measurement of education outcomes (editorial). *Journal of the Royal Statistical Society*, A, 155, 313–15.

Goldstein, H. and McDonald, R.P. 1987: A general model for the analysis of multilevel data. *Psychometrika* 53, 455–67.

Goldstein, H. and Rasbash, J. 1992: Efficient computational procedures for the estimation of parameters in multilevel models based on iterative generalised least squares. *Computational Statistics and Data Analysis* 13, 63–71.

Goldstein, H., and Healy, M.J.R. 1994: The graphical presentation of a collection of means. *Journal of the Royal Statistical Society* A, 157 (to appear).

Goldstein, H., Healy, M.J.R. and Rasbash, J. 1994: Multilevel time series models with applications to repeated measures data. *Statistics in Medicine* 13, 1643–55.

Goldstein, H., Rasbash, J., Yang, M., Woodhouse, G., Pan, H., Nuttall, D. and Thomas, S. 1993: A multilevel analysis of school examination results. *Oxford Review of Education* 19, 425–33.

Greenacre, M.J. 1984: *Theory and applications of correspondence analysis.* New York: Academic Press.

Grizzle, J.C. and Allen, D.M. 1969: An analysis of growth and dose response curves. *Biometrics* 25, 357–61.

Gumpertz, M.L. and Pantula, S.G. 1992: Nonlinear regression with variance components. *Journal of the American Statistical Association* 87, 201–9.

Harrison, G.A. and Brush, G. 1990: On correlations between adjacent velocities and accelerations in longitudinal growth data. *Annals of Human Biology* 17, 55–7.

Heath, A., Jowell, R., Curtice, J., Evans, G., Field, J. and Witherspoon, S. 1991: *Understanding political change.* Oxford: Pergamon.

Hedges, L.V. and Olkin, I.O. 1985: *Statistical methods for meta analysis.* Orlando, Florida: Academic Press.

Holland, P.W. 1986: Statistics and causal inference. *Journal of the American Statistical Association* 81, 945–71.

Jenss, R.M. and Bayley, N. 1937: A mathematical method for studying the growth of a child. *Human Biology* 9, 556–63.

Joreskog, K.G. and Sorbom, D. 1979: *Advances in factor analysis and structural equation models.* Cambridge, MA: Abt Books.

Kreft, I.G., de Leeuw, J. and van der Leeden, R. 1994: Comparing five different statistical packages for hierarchical linear regression: BMDP–5V, GENMOD, HLM, ML3, and VARCL. *American Statistician* 48 (to appear).

Laird, N.M. and Louis, T.A. 1987: Empirical Bayes confidence intervals based on bootstrap samples. *Journal of the American Statistical Association* 82, 739–57.

Laird, N.M. and Louis, T.A. 1989: Empirical Bayes confidence intervals for a series of related experiments. *Biometrics* 45, 481–95.

Larsen, U. and Vaupel, J.W. 1993: Hutterite fecundability by age and parity: strategies for frailty modelling of event histories. *Demography* 30, 81–101.

Lawley, D.N. and Maxwell, A.E. (1971). *Factor analysis as a statistical method* (2nd edition). London: Butterworth.

Liang, K. and Zeger, S.L. 1986: Longitudinal data analysis using generalised linear models. *Biometrika* 73, 45–51.

Lindley, D.V. and Smith, A.F.M. 1972: Bayes estimates for the linear model. *Journal of the Royal Statistical Society*, B 34, 1–41.

Lindstrom, M.J. and Bates, D.M. 1990: Nonlinear mixed effects models for repeated measures data. *Biometrics* 46, 673–87.

Little, R.J.A. 1992: Regression with missing X's: a review. *Journal of the American Statistical Association* 87, 1227–37.

Longford, N.T. 1987: A fast scoring algorithm for maximum likelihood estimation in unbalanced mixed models with nested random effects. *Biometrika* 74, 817–27.

Longford, N.T. 1988: VARCL – software for variance component analysis of data with hierarchically nested random effects (maximum likelihood). Princeton, NJ: Educational Testing Service.

Longford, N.T. 1993: *Random coefficient models*. Oxford: Clarendon Press.

Longford, N.T. and Muthen, B.O. 1992: Factor analysis for clustered populations. *Psychometrika* 57, 581–97.

McCullagh, P. and Nelder, J. 1989: *Generalised linear models* (2nd edition). London: Chapman and Hall.

McDonald, R.P. 1985: *Factor analysis and related methods*. Hillsdale, New York: Lawrence Erlbaum.

McDonald, R.P. 1993: A general model for two level data with responses missing at random. *Psychometrika* 58, 575–85.

McDonald, R.P. 1994: The bilevel reticular action model for path analysis with latent variables. *Sociological Methods and Research* 22, 399–413.

McDonald, R.P. and Goldstein, H. 1988: Balanced versus unbalanced designs for linear structural relations in two level data. *British Journal of Mathematical and Statistical Psychology* 42, 215–32.

McGrath, K. and Waterton, J. (1986). British Social Attitudes, 1983–1986 panel survey. London: Social and Community Planning Research.

Mason, W.M., Anderson, A.F. and Hayat, N. 1988: *Manual for GENMOD*. Ann Arbor: University of Michigan Population Studies Centre.

Miller, R.G. 1974: The Jacknife – a review. *Biometrika* 61, 1–15.

Mortimore, P., Sammons, P., Stoll, L., Lewis, D. and Ecob, R. 1988: *School matters*. Wells: Open Books.

Moulton, L.H. and Zeger, S.L. 1989: Analysing repeated measures on generalised linear models via the bootstrap. *Biometrics* 45, 381–94.

Muthen, B.O. 1989: Latent variable modelling in heterogeneous populations. *Psychometrika* 54, 557–85.

Nuttall, D.L., Goldstein, H., Prosser, R. and Rasbash, J. 1989: Differential school effectiveness. *International Journal of Educational Research* 13, 769–76.

Paterson, L. 1991: Socio-economic status and educational attainment: a multidimensional and multilevel study. *Evaluation and Research in Education* 5, 97–121.

Peto, R. 1972: Contribution to discussion of paper by D.R. Cox. *Journal of the Royal Statistical Society*, B 34, 205–7.

Plewis, I. 1985: *Analysing change.* Chichester: Wiley.

Plewis, I. 1993: Reading progress. In G.Woodhouse (ed.) *A Guide to ML3 for new users.* London: Multilevel Models Project.

Plewis, I. 1994: Statistical methods for understanding cognitive growth: a review, a synthesis and an application (In press).

Prosser, R., Rasbash, J. and Goldstein, H. 1991: *ML3 Software for Three-level analysis: user's guide for version 2.* London: Institute of Education.

Rasbash, J., Yang, M., Woodhouse, G. and Goldstein, H. 1995: *Mln: command reference guide.* London: Institute of Education.

Raudenbush, S.W. 1993: A crossed random effects model for unbalanced data with applications in cross-sectional and longitudinal research. *Journal of Educational Statistics* 18, 321–49.

Raudenbush, S.W. 1994: Equivalence of Fisher scoring to iterative generalised least squares in the normal case with application to hierarchical linear models. Unpublished.

Robinson,W.S. 1950: Ecological correlations and the behaviour of individuals. *American Sociology Review* 15, 351–7.

Rosier, M.J. 1987: The second international science study. *Comparative Education Review* 31, 106–28.

Royall, R.M. 1986: Model robust confidence intervals using maximum likelihood estimators. *International Statistical Review* 54, 221–6.

Rubin, D.B. 1987: *Multiple imputation for nonresponse in surveys.* New York: Wiley.

Searle, S.R., Casella, G. and McCulloch, C.E. 1992: *Variance components.* New York: Wiley.

Seltzer, M.H. 1993: Sensitivity analysis for fixed effects in the hierarchicl model: a Gibbs sampling approach. *Journal of Educational Statistics* 18, 207–36.

Skinner, C.J., Holt, D. and Smith, T.M.F. 1989: *Analysis of complex surveys.* Chichester: Wiley.

Snijders, T.A.B. and Bosker, R.J. 1993: Standard errors and sample sizes for two-level research. *Journal of Educational Statistics* 18, 237–59.

Vevea, J. 1994: A model for estimating effect size in the presence of publication bias. Paper presented to American Educational Research Association, annual meeting, New Orleans, April 1994.

Waclawiw, M.A. and Liang, K. 1993: Prediction of random effects in the generalised linear model. *Journal of the American Statistical Association* 88, 171–8.

Waclawiw, M.A. and Liang, K. 1994: Empirical Bayes estimation and inference for the random effects model with binary response. *Statistics in Medicine* 13, 541–51.

Wei, L.J., Lin, D.Y. and Weissfeld, L. 1989: Regression analysis of multivariate incomplete failure time data by modelling marginal distributions. *Journal of the American Statistical Association* 84, 1065–73.

Wolfinger, R. 1993: Laplace's approximation for nonlinear mixed models. *Biometrika* 80, 791–5.

Woodhouse, G., Yang, M., Goldstein, H., Rasbash, J. and Pan, H. 1995: Adjusting for measurement error in multilevel analysis. *Journal of Educational and Behavioural Statistics* (to appear).

Zeger, S.L. and Karim, M.R. 1991: Generalised linear models with random effects; a Gibbs Sampling approach. *Journal of the American Statistical Society* 86, 79–102.

Zeger, S.L., Liang, K.-Y. and Albert, P.S. 1988: Models for longitudinal data: a generalised estimating equation approach. *Biometrics* 44, 1049–60.

Author Index

Subject Index

abortion 55–7, 60, 97
accelerated life model *see* log duration model
additive model for variance 119
aggregate level variable 160
assumptions in model 27–8
autocorrelation; *see* time series

Bayesian linear model 23, 24
bias 62, 66, 126
binary response data 11, 62, 85, 109, 110
binomial distribution 86, 97–9, 102, 103, 107–11, 132, 134
binomial–extra binomial distribution 99
BIRAM 153, 161, 163
birth interval 130, 131, 134, 136, 137
birthweight 88, 94
biserial correlation (covariance) 110, 135
bivariate model 94, 105, 110, 121, 122, 134, 136, 161
blocking factor 129, 130, 132
BMDP 153, 163
bone age 89, 90, 91
bootstrap 36, 60–3
bootstrap *see also* nonparametric, parametric
British Election Study 101
BUGS 153, 163

categorical variable *see* multicategory variable

causality 11
ceiling effect 49
censored data 125–7, 133, 134, 135
    left 134
    right 126–7, 134, 136, 137
census 10
centring 27, 30
class size 11
cluster 1, 3, 5, 6, 8, 9
coefficient of variation 50, 53, 94
competing risk 132
complete data 158, 74
completed data matrix 155–8, 159
complex level 1 variance *see* complex variation
complex variation 21, 47, 48–57, 64, 133, 154
    level 1 48, 55, 80, 134
compositional effect or measurement 8, 9, 18, 30–2, 49, 65
computational efficiency 61, 73, 74, 90, 154, 162
computing time 118
confidence interval 1, 3, 5, 21, 22, 26, 32–7, 42, 60–3, 86, 101
    bootstrap 62
    overlapping 36
    residuals 36
confidence region 34, 35
    see also confidence interval
confounding 11, 12, 120

Books are to be returned on or before
the last date below.

2468

- 3 NOV 199

- 8 APR 2005